W9-AVG-409

Rice Policy in Indonesia

Rice Policy in Indonesia

Scott Pearson • Walter Falcon
Paul Heytens • Eric Monke
Rosamund Naylor

CORNELL UNIVERSITY PRESS
Ithaca and London

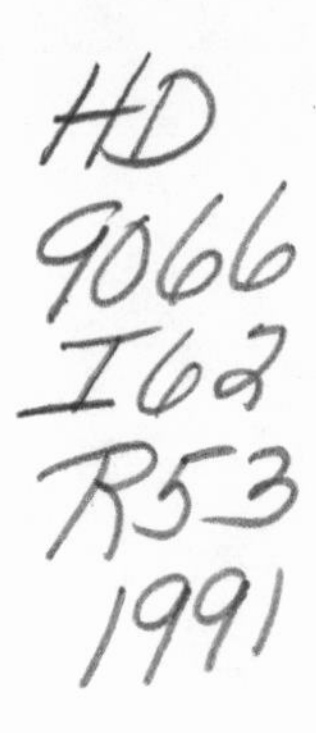

Cornell University Press gratefully acknowledges a grant from Stanford University which aided in bringing this book to publication.

First published 1991 by Cornell University Press.

International Standard Book Number 0–8014–2524–7
Library of Congress Catalog Card Number 90-55751
Printed in the United States of America
Librarians: Library of Congress cataloging information appears on the last page of the book.
♾ The paper in this book meets the minimum requirements of the American National Standard for Information Sciences—Permanence of Paper for Printed Library Materials, ANSI Z39.48–1984.

Contents

Foreword

Sjarifudin Baharsjah
Junior Minister, Ministry of Agriculture, Republic of Indonesia

The Food Research Institute at Stanford University carried out three in-depth studies of Indonesia's principal food crops—cassava, corn, and rice—during the 1980s. This work has served as an invaluable complement to our own efforts at analyzing food and agricultural development policy during a period that witnessed Indonesia's transformation from a major rice importer into a self-sufficient producer.

During my long acquaintance with the Stanford research teams, I have been impressed by their modus operandi because of their effective collaboration with diverse institutions and the methodological rigor with which they pursue their multidimensional analysis of single commodities. Most significant, their comprehensive analytical framework for food policy analysis stresses the importance of assessing the actual or potential trade-offs between the national objectives of efficient income growth, equitable distribution of income, and security of food supplies for consumers.

In this spirit, the current volume traces the fortunate circumstances of minor trade-offs in rice policy through the time of the transition from rice imports to self-sufficiency. The cautionary note regarding the adverse efficiency and equity impacts of improving future food security by means of substantial increases in domestic food prices is also well founded.

To policymakers, the conclusion that as rice output more than doubled, aggregate employment in rice production was unchanged is sobering. The revelation that the real wages of unskilled agricultural workers increased substantially during the 1980s is both an important cause of rice farmers' decisions to employ much less labor per unit of output and a key indicator of the improving living standards of Indonesia's rural poor. Finally, a detailed analysis of the changing structure of the world market for rice shows the declining importance of Asian countries as importers but illustrates that the world rice price will probably continue to be very unstable and difficult to predict.

Empirical insights such as these, drawn within an integrated analytical framework, provide a reliable evaluation of past policy and serve as guidelines in the design and implementation of future policy. I rest confident that the Food Research Institute will continue its collaboration in superior food and agricultural policy analysis as Indonesia develops further.

Acknowledgments

The Stanford Project team received an enormous amount of assistance while conducting this study. Our research was sponsored by the U.S. Agency for International Development and the government of Indonesia under a contract between BAPPENAS and Stanford University titled "Food Policy and Rural Income Generation," dated July 24, 1986. Special thanks are owed to numerous individuals at all levels in the team's counterpart agencies in the Indonesian government—BAPPENAS, BULOG, and the Ministry of Agriculture—and to officials in the central, regional, and local offices of the Central Bureau of Statistics and the Ministry of Manpower, who collectively made the fieldwork possible. Critical leadership was provided by Minister Saleh Affif, Minister Bustanil Arifin, Junior Minister Sjarifudin Baharsjah, General Sukriya Atmaja, and Soetatwo Hadiwigeno. Several professionals in the U.S. AID Mission to Indonesia offered valuable substantive and administrative help. Many Indonesian farmers, employers, and workers gave freely of their time and experiences. At Stanford, Will Masters offered invaluable research assistance, contributing centrally to the first and last chapters of the book, Claudia Smith provided expert secretarial skills, and Susan Maher looked after all administrative details, especially those of a budgetary nature. To all of these institutions and individuals—and to two anonymous referees—we offer an enthusiastic vote of thanks.

Abbreviations and Indonesian Terms

Government Organizations and Programs

BAPPENAS	Badan Percencanaan Pembangunan Nasional (national planning agency)
BIMAS	Bimbingan Masai ("mass guidance" intensification program)
BPS	Biro Pusat Statistik (Central Bureau of Statistics)
BULOG	Badan Urusan Logistik (National Food Logistics Agency)
KUD	Koperasi Unit Desa (village cooperative)
KUPEDES	Kredit Umum Pedesaan (rural credit program)
REPELITA	Rencana Pembanungan Lima Tahun (five-year development plan)
SUSENAS	Survei Sosial Ekonomi Nasional (national social economic survey)
TRI	Tebu Rakyat Intensifikasi (small-holder sugarcane intensification program)

Indonesian Terms

Ani-ani	Small hand knife to cut the rice stalk
Bawon	Traditional harvest arrangement in which harvesters earn a fixed share of their output
Becak	Rickshaw
Ceblokan	Labor-hiring arrangement for planting in which workers earn special rights to the harvest
Gabah	Threshed paddy
Gotong-royong	Mutual exchange of unpaid labor
Kabupaten	District
Kedokan	Labor-hiring arrangement for preharvest tasks in which workers earn special rights to the harvest
Merantau	Extended migration to a distant region for work
Padi gogo rancah	Upland dry-seeded rice
Palawija	Nonrice staples or secondary crops
Penebas	Entrepreneur who organizes *tebasan* harvests
Rupiah	Indonesian currency
Sawah	Wet rice field
Tebasan	A system for harvesting in which the farmer sells his or her standing crop of paddy to a middleman (*penebas*)

1. Introduction

Scott Pearson and Eric Monke

In 1991, agricultural policymakers in Indonesia confront a frustrating dilemma. Virtually all the main economic indicators—positive contribution to rural income and employment, efficient saving of foreign exchange, and security of domestic food supplies—point to the desirability of expanding domestic rice production to keep pace with the growth of Indonesian consumption. Yet all available policy options to induce farmers to grow more rice are severely constrained. Greater public investments in irrigation and research are limited by increased competition for government funds after several years of budgetary stringency, higher rice prices are unpopular with rice consumers and macro policymakers concerned with inflation, and conversion of irrigated sugar land into crop rotations involving two or three annual crops of rice runs headlong into government efforts to utilize sugar refining capacity and reduce imports of sugar.

This book examines the components of Indonesia's rice policy dilemma to help policymakers, analysts, and observers sort out the pros and cons of alternative courses of action. The study combines new field-based empirical evidence on the profitability of rice farming and on rural employment and wages with long experience in the analysis of food policy issues in Indonesia and the international market for rice.[1] The intended result is policy-relevant information gleaned from microeconomic studies of rice production systems and rural labor markets set within a macro food policy context.

[1]This book contains the principal results of the Food Research Institute's third multi-year research project on Indonesian food policy within the past decade. The results of the first two projects are reported in Falcon, Jones, Pearson, et al. (1984), and in Timmer, ed. (1987). The conceptual framework for all three investigations is presented in detail in Timmer, Falcon, and Pearson (1983). The methodology underlying the analysis of comparative advantage and policy transfers in Indonesia's rice production systems, presented in Chapter 7 of this volume, is set forth in Monke and Pearson (1989).

This analysis of Indonesia's rice economy draws on the earlier analysis and insights of two colleagues, Leon A. Mears and C. Peter Timmer. The authors also have benefited greatly from having access to work in progress of several agricultural economists working in the Ministry of Agriculture, most notably Klaus Altemeier, Faisal Kasryno, and Steven Tabor. A recent book, presenting contrasting views, is Booth (1988).

Framework for Agricultural Policy Analysis

Alternative Strategies

The Indonesian government faces a choice among three broad strategies for rice policy. The first possible direction is to aim for a rapid acceleration of growth in rice production, to raise and keep annual output well above consumption levels. This effort would allow absolute self-sufficiency, ensuring that the country would never import rice and in most years would have rice supplies available for export. At current levels of growth in consumption, this strategy would involve expanding rice production at perhaps 4 percent per year over the next five years.

At the opposite extreme, the country could aim to diversify the rural economy, encouraging rice farmers to grow a range of other crops and undertake off-farm activities. This strategy of diversification would require frequent and increasingly large rice imports, although local surpluses might be available in years of exceptionally good harvests. On average, rice production might grow only 1 percent annually, with much more rapid growth in other income-earning activities.

An intermediate strategy would target production growth roughly equal to consumption growth, at about 2.5 percent per year over the next half decade. To maintain steady consumption levels at stable prices, this strategy would entail either considerable storage of rice from good years into bad, occasional imports and exports, or some combination of the two. Even with frequent rice trades, however, on average exports would equal imports and the country would follow a trend of self-sufficiency.

This book discusses what would be required to reach each of these alternative target growth rates and what effects the pursuit of each strategy would have on the Indonesian economy. It focuses on the impact of rice policy on rural income and employment but also considers its effects on urban consumers through the level and stability of rice prices and on government expenditures through subsidized storage and trade. Although no definitive judgments about relative priorities are made, whenever possible the advantages and disadvantages of the three alternative strategies are spelled out and quantified.

Available Instruments

Pursuing any of these strategies—high growth with exports, low growth with imports, or trend self-sufficiency—requires the use of several interrelated policy instruments. These instruments can be grouped into five categories: price level policies, price stabilization policies, public investment policies, macroeconomic policies, and crop regulatory policies.

For rice in Indonesia, the two main price levels are those of rice and fertilizer; each year the government sets and defends a floor price for unmilled rice (*gabah*) and a set of wholesale prices for the major types of chemical fertilizers. The National Food Logistics Agency (BULOG) defends the rice floor price through government purchases of *gabah* and milled rice. BULOG also works to stabilize the consumer price of rice by injecting stored or imported rice supplies into domestic markets when shortages cause the price to rise excessively. Accomplishing these tasks has required large and ongoing subsidies through the credit system and from tax revenues, but BULOG has been successful in defending appropriate rice prices during the past fifteen years. Although price stability has been an important element of Indonesia's success in expanding rice production, specific stabilization mechanisms are not discussed at length in this book.[2]

The major public investments (capital spending from the government budget) affecting rice production are the construction and maintenance of infrastructure for irrigation and transportation and the provision of research and extension services. Expenditures in these areas have had very high payoffs. But the highest rate of investment was reached when government oil revenues were large, and the rate of spending has fluctuated along with oil income. Growth in the area planted to high-yielding irrigated rice varieties has been particularly dependent on these investments.

Macroeconomic policies include government revenues and expenditures (fiscal policy), the rate of expansion of the money supply and hence the cost and availability of credit (monetary policy), and the exchange rate for Indonesia's rupiah relative to foreign currencies (exchange rate policy). These policies are set for the economy as a whole, but they have a profound effect on agricultural investment and growth because they affect the profitability of agricultural production. Unlike those in many other developing countries, Indonesian macroeconomic policies generally have been balanced across sectors, avoiding heavy taxation of the agricultural sector and promoting sustainable growth throughout the economy.

The final category of policy instruments includes rural production programs and regulations, which directly influence farm activities in certain areas. In Indonesia, rice hectarage has been reduced because of government-enforced production targets for sugarcane, tobacco, and, more recently, soybeans. The degree of enforcement and the competitiveness with the rice production calendar vary among crops and locations but seem most significant for sugarcane in East and Central Java. The desire to promote self-sufficiency in sugar and to utilize capacity in the sugar

[2]BULOG's price stabilization activities are analyzed in Falcon et al. (1985) and in Timmer (1989), pp. 22–64.

processing industry has led to the enforced use of about 150,000 hectares of highly productive, irrigated land for sugarcane.

Governments typically enact policies individually and often do not consider all of the interacting influences that policies have on one another. In part, this occurs because of the division of responsibility among government ministries and agencies. Nevertheless, it is desirable analytically for planners to consider the strategies implied by individual or collective policy decisions. Each of these policy instruments depends to a considerable degree on the others for its success or failure. For example, rural irrigation infrastructure may be necessary for fertilizer applications to increase yields reliably, while rice price policy may determine how much fertilizer it is profitable for farmers to use in any given year.

Because policies tend to reinforce each other so strongly, the approach taken in this book is to analyze the contributions of each category of the various policy instruments to rice output, farm income, and rural income distribution. Then the policies are viewed in the context of agricultural strategy to learn whether collectively they can deliver the quantities of additional rice output needed to fulfill each strategy. In this way, the analysis contrasts the three alternative targets of high growth (and absolute rice self-sufficiency), low growth (and rural diversification), and medium growth (for trend self-sufficiency). Each of these targets can be achieved with a range of policy instruments, but each strategy has advantages and disadvantages.

Government Objectives

The appropriateness of the individual and collective policies underlying each of the three strategies considered here can be evaluated by assessing how well they advance government objectives. Three fundamental indicators of national welfare—food security and price stability, rapid income growth, and desirable income distribution—can be seen as the central objectives of government policy.[3]

The first objective, food security at the national level, is the capability of the country to produce adequate amounts of foodstuffs for all consumers at affordable prices. Since food shortages are quickly reflected in rising food prices, food security is closely related to the government's ability to maintain stable domestic food prices.

The second objective, income growth, demands that resources be allocated as efficiently as possible. National income will be maximized if land, labor, and all other goods and services in the economy are allocated to their most profitable uses, when profits are measured in terms of the

[3]This conceptual approach is spelled out in the first chapter of Timmer, Falcon, and Pearson (1983), pp. 3–18.

social value or opportunity cost of each good. This goal is common to government policy-making in all sectors of the economy. It implies that policies should create incentives for individuals to make the most valuable use of the nation's resources in both production and consumption.

The third objective of food policy is a satisfactory distribution of income to needy groups and to poorer regions. Because the production incentives of food policy apply to the agricultural sector, the income distributional objective often refers to greater incomes being generated for the rural poor, including farmers, agricultural laborers, and rural workers employed off the farm. Since food policy influences consumer prices for food and other agricultural goods, the real incomes of the urban poor are affected as well. Issues involving regional income distribution go beyond the consideration of urban versus rural inequities. The economic importance of agriculture in many of the poorer regions of Indonesia means that food policies inevitably influence the welfare of all groups; higher rice prices, for example, could transfer income from poor rural rice consumers to relatively well-off rice farmers, many of them living in higher-income regions.

These three government objectives are closely interrelated. At times they may be in conflict with one another so that the government must trade off one objective for another. For example, holding a large food security rice stock ties up government funds that could be invested elsewhere and therefore reduces the growth rate of the economy as a whole. At other times, these objectives may be complementary because two or three of them are served simultaneously by a given policy. Investment in rural infrastructure, for example, can both generate growth and improve income distribution. An important element of the analysis in this book, therefore, is to link the effects of policy to all three fundamental food policy objectives, looking for a rice strategy that will exploit potential complementarities among them.

The analysis in this book brings together the linkages among rice policy instruments, rice-related variables in the rural economy, and the objectives of government to answer the question, Does more rice output (good for food security) create efficient income growth (required for efficiency) and labor-intensive employment (needed for better income distribution)? The study evaluates the effectiveness of past rice policy and the options for future policy by tracing empirically the relationships among five categories of policy instruments (price level, price stabilization, public investment, macro, and crop regulatory), three principal economic variables (rice output, income, and employment), and three fundamental food policy objectives (food security, efficiency, and income distribution). The links between policy and the economy can be traced conveniently through Figure 1.1, which shows how a given strategy, implemented

FIGURE 1.1. *Linkages among rice strategy, policy instruments, the economy, and government objectives*

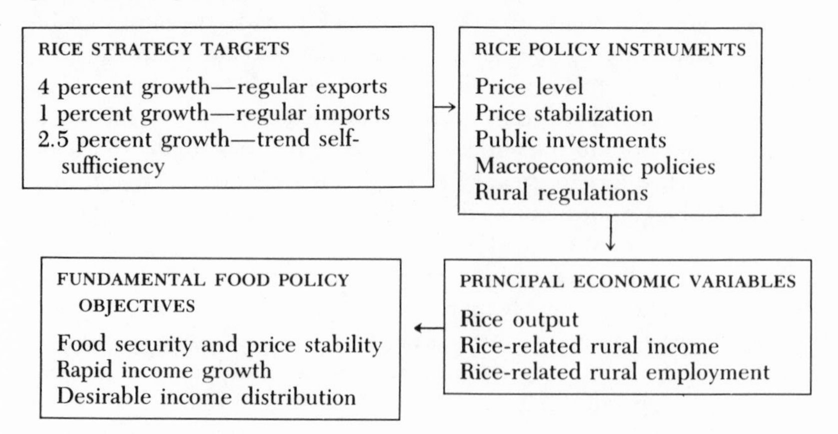

through a variety of policy instruments, affects several economic variables that might further the government's objectives, thus conditioning the government's choice of strategy.

Organization and Content of the Book

This book is divided into three sections, plus this introduction, which outlines the analytical and historical context of the study, and a conclusion, which summarizes the study and its principal results.

The first section (Chapters 2 and 3) provides the analytical foundation for the book, drawing links among food policy objectives, rice policies, and the resulting levels of rice output, income, and employment. Chapter 2 is a survey of the recent history of all five categories of policy that influence the production and price of rice. Some of these policies are also the instruments by which Indonesia escapes full dependence on the world market, helping to reduce the level and instability of domestic prices. In Chapter 3, the external market for Indonesian rice is reviewed. The world price is the starting point for rice policy because Indonesians almost certainly would trade at that price if no policy were in place. The chapter shows how the price of Indonesian imports or exports of rice is determined, why the world rice price is inherently unstable, and why predictions of future world prices are so difficult to make with much accuracy.

The second section (Chapters 4, 5, and 6) covers the current situation and recent changes in output, income, and employment levels in Indonesia's main rice production systems. It documents the successes of the policies enacted during the 1970s and 80s through support of *gabah* prices and use of public investment to raise domestic rice output and stabilize domestic rice prices. Chapter 4 contains an empirical examination, based on original fieldwork, of the profitability of Indonesia's main wetland rice production systems in 1987. This chapter gives a detailed cross section of technological practices and income levels from rice. Then, in Chapter 5, data on rural labor markets document changes in employment opportunities, labor market institutions, and real wages during the 1980s. The aim is to investigate the competitiveness of rural labor markets and the trends in real rural wages to aid later analysis of the efficiency and equity effects of alternative rice strategies. In Chapter 6, time series data are combined with backward extrapolations of the 1987 budgets (to 1980 and 1969) to investigate how input substitution and technical change in rice farming have affected the expansion of rice output, employment, and rice farming profits on Java.

The third and last section of this book goes to the heart of the inquiry: what direct and indirect effects might future rice policies have on the level and distribution of rural income? The first part of this question is addressed in Chapter 7, which presents the production and profitability effects of alternative sets of policies that might underlie the strategies. The second is addressed in Chapter 8, which describes the impact of those policies on rural employment, wages, and income distribution.

The book ends with a set of conclusions, reviewing the impact of policy in the past and the prospects for the future. Although the spectacular increases in rice productivity of the recent past might not be achievable again, sustained improvements in the level and distribution of rural income are clearly possible. But this desired outcome will eventuate only if policymakers are successful in finding an appropriate set of rice policies. Resourceful, innovative farmers remain ready to expand rice production in response to public incentives. In question is whether the political consensus can be found to put in place an appropriate strategy.

2. Recent Policy Influences on Rice Production

Scott Pearson, Rosamund Naylor, and Walter Falcon

This chapter reviews the complex set of relationships among rice strategies, policy instruments, economic variables, and food policy objectives as they have evolved since the early 1970s. The intent is to provide readers unfamiliar with the recent history of Indonesian rice policy with a summary of rice policy and production performance and to offer already knowledgeable readers an interpretive approach to understanding how strategies, policies, variables, and objectives have fit together. The modest goal, then, is to review the set of policy instruments available to Indonesian decision makers, whereas the more ambitious intention is to interpret these instruments' recent use in the context of food policy objectives.[1]

Recent Rice Policy in the Context of Food Policy Objectives

The chain of reasoning summarized here offers readers a set of guideposts as they work their way through the details of policy that follow. In response to the highly unstable world price of rice and detrimental short-run influences on the world price of Indonesia's decisions to buy or sell rice on the world market (described in Chapter 3), Indonesian policymakers in all of the New Order governments have given central place to the food security objective within their rice strategy of self-sufficiency. The desire for food security became even stronger during 1973–75, when world prices of rice reached historically unprecedented heights. To ensure that most consumers would have access to rice supplies at affordable prices, the government has carried out policies to influence two related variables in the rice economy—the stability of domestic rice prices and the level of Indonesian rice production. Food security thus was served by

[1]C. Peter Timmer recently summarized his own work as well as that of others in an analysis of two decades of Indonesian food policy; see his "Indonesia: Transition from Food Importer to Exporter," in Sicular, ed. (1989), pp. 22–64. That essay also has a complete bibliography (pp. 61–62), which includes references to important earlier articles on the Indonesian rice economy such as Afiff and Timmer (1971), and Afiff, Falcon, and Timmer (1980). Readers who wish to compare Indonesia's rice policy with those of other Asian countries are encouraged to consult Sicular, ed. (1989) and Barker, Herdt, and Rose (1985).

enacting price stabilization policies, involving public storage of rice and imports, and by creating incentives for expanded domestic rice production through the use of price level policy (guaranteed minimum rice prices to farmers and subsidies on fertilizers) and public investment policy (irrigation infrastructure and maintenance, transportation facilities, and research, development, and dissemination of seeds and technologies for high-yielding varieties [HYVs]).

Beginning in 1973–74, when oil prices surged upward, Indonesia enjoyed rapid economic growth and burgeoning government revenues, which permitted policymakers to give increasing attention to the equity objective of food policy, in particular to the desirability of increasing incomes of the rural poor. They saw that the uneven distribution of the oil boom income growth, which favored urban residents, could be offset at least in part by carrying out agricultural policies that encouraged generation of income and employment in rural areas. Price and investment policies that were intended principally to enhance food security by increasing domestic supplies of food also had the desirable side effect of redressing the imbalance of skewed income distribution between rural and urban areas. No trade-offs were necessary because both the objectives of food security and of rural-urban income distribution were furthered by policies that led to increased rice output.

In the early 1980s, the world oil price began to slide downward and by the middle of the decade had settled in a range less than half of its 1980 peak. With the end of the oil boom, the Indonesian economy slipped into slow growth and entered a difficult period of macroeconomic adjustment. Accordingly, policymakers redoubled their efforts to find activities in the economy to enable income growth to occur efficiently, with less reliance on government budgetary expenditures. Agriculture, and especially rice production, became recognized as a prime source of efficient growth. Hence a third basic objective of food policy, efficient income growth, came to the fore of Indonesian rice policy. Once again, to the extent that rice (substituting for imports) could be produced efficiently, complementarity occurred among the objectives of food security, rural-urban equity, and efficient income growth; all could be advanced by policies that spurred growth of rice production.

The achievement of rice self-sufficiency (zero imports) in 1985 ended, at least temporarily, this happy complementarity among the three fundamental food policy objectives. Whereas income distribution favoring rural over urban areas would continue to be furthered by policies that promoted increasing rice production, any excess of domestic production over consumption would have to be exported or stored. Public storage beyond levels needed for food reserve stocks did little to improve food security. Instead, surplus stocks represented burdensome costs for an already strapped budget. Exports were costly as well because they required public subsidies to enter the world market. Whereas rice was competitive against the c.i.f. price of imports (with occasional implicit subsidies from consumers), domestic prices were well above f.o.b. prices for exports. Ironically, the attainment of self-sufficiency highlighted the inevitably close links between Indonesian rice policy and the world rice market. Rice supplies in excess of consumption needs did not create efficient growth of income, were not an appropriate vehicle for food security because they were too expensive to store in large quantities, but were still desired for improved income distribution in rural areas. Indonesian policymakers in the mid-1980s thus faced a classic food policy trade-off in assessing the strategic alternatives for rice production.

Since 1985, the rice economy has been at or near self-sufficiency. As shown in Chapter 3, the world rice price went up about 50 percent in 1987, slipped back in 1988, and rose and fell again in 1989. Indonesian rice prices were raised in late 1987 and 1988 to stay in line with world prices and were about aligned with world prices in late 1989. Excess Indonesian public stocks were either exported with subsidy (in 1985–86) or injected into the domestic market to replace unexpected production shortfalls (in 1987–88). Food reserve stocks were depleted in order to balance domestic supply and demand at higher prices during times of production shortfalls (1987–88) and then rebuilt when production recovered (1989).

Within only four years, Indonesia fluctuated from a situation of too much rice to one of too little to one of comfortable adequacy. As a result, new trade-offs have emerged among the basic objectives of food policy. Should Indonesia transfer scarce budgetary and consumer resources into rice production to avoid imports to the greatest possible extent (absolute self-sufficiency), invest enough so that production and consumption normally are in balance, using imports or exports to offset occasional gaps (trend self-sufficiency), or diversify out of rice and import as required (regular imports)?

Recent History of Policy Influences on Rice Production

Rice self-sufficiency and stable rice prices have been the focus of agricultural policy in Indonesia throughout the past two decades. The strategy adopted by the government in the late 1960s to further these ends had two main policy components. The first was a comprehensive set of price policies that controlled the level and stability of consumer and producer prices of rice as well as the prices of important inputs such as fertilizer and pesticides. The second was public investment in modern rice-producing technologies and in the infrastructure and institutions needed to support the new technologies. The instruments used within each of these policy categories are discussed in the remainder of this chapter. To provide historical context on the evolution and impact of policy, information on rice output, income, and employment is summarized at the outset of this section.

Recent Performance of the Indonesian Rice Economy

Between 1955 and 1965, Indonesia had the lowest rate of growth in rice yields (0.2 percent per year) and rice production (1.2 percent) of any major rice producer. Between 1965 and 1985, it had the highest rates of growth (4.1 percent and 5.6 percent, respectively), with a dramatic spurt of 7.2 percent annual growth of output between 1977 and 1984. As shown in Table 2.1, rice production in Indonesia grew by nearly two and one-half times between 1968 and 1989, from less than 12 to over 29 million metric tons. Most of this expansion occurred during the second of these two decades, when average paddy (*gabah*) yields increased from 2.8 to 4.2

TABLE 2.1. *Indonesian rice data, 1968–90*

Year	Area (million hectares)	Paddy (*gabah*) production (million metric tons)	Rice production (million metric tons)	Paddy yield (tons per hectare)	Paddy support price (rupiahs per kilogram)	Urea price (rupiahs per kilogram)	Ratio of paddy price to urea price
1968	8.01	17.16	11.67	2.14	23	40	0.58
1969	8.01	18.02	12.25	2.25	21	32	0.66
1970	8.14	19.33	13.14	2.38	18	27	0.67
1971	8.32	20.19	13.72	2.43	21	27	0.78
1972	7.90	19.39	13.18	2.45	21	26	0.81
1973	8.40	21.49	14.61	2.56	30	40	0.75
1974	8.52	22.47	15.22	2.64	42	40	1.05
1975	8.50	22.34	15.18	2.63	59	60	0.98
1976	8.37	33.30	15.84	2.78	69	80	0.86
1977	8.36	23.35	15.88	2.79	71	70	1.01
1978	8.93	25.77	17.53	2.89	75	70	1.07
1979	8.80	26.28	18.87	2.99	85	70	1.21
1980	9.01	29.65	20.16	3.29	105	70	1.50
1981	9.38	32.77	22.29	3.49	120	70	1.71
1982	9.90	33.58	22.84	3.39	135	70	1.93
1983	9.16	35.30	24.01	3.85	145	90	1.61
1984	9.76	38.14	25.93	3.91	165	90	1.83
1985	9.90	39.03	26.54	3.94	175	100	1.75
1986	9.90	39.39	26.78	3.98	175	125	1.40
1987	9.92	40.08	27.25	4.04	190	125	1.52
1988	10.14	41.67	28.33	4.11	210	135	1.56
1989	10.21[a]	43.21[a]	29.38[a]	4.23[a]	250	165	1.52
1990	NA	NA	NA	NA	270	185	1.46

SOURCE: BULOG.
NA = Not available.
[a]Preliminary estimates.

TABLE 2.2. *Structure of the agricultural economy, 1968–87 (percentage shares)*

	1968	1975	1980	1983	1985	1987
Agriculture in GDP at current market prices	51.0	31.7	24.8	24.0	23.5	25.5
Food crops in agriculture	67.9	63.8	56.3	62.5	62.4	60.4
Rice in food crops	54.4	51.2	49.0	46.3	48.3	52.5
Rice in agriculture	37.0	32.7	27.6	28.9	30.1	31.7
Rice in GDP	18.8	10.3	6.8	6.9	7.1	8.1

SOURCE: BPS.

tons per hectare. Area planted to rice expanded by only about 1 million hectares during each decade, and so most of the output gain was attributed to intensive productivity increases rather than to extensive expansion of rice land. These production data provide evidence of the broad success of Indonesian rice policy in encouraging growth of rice output.

During this same twenty-year period, the structure of the entire Indonesian economy changed markedly. Table 2.2 illustrates the shift in the agricultural sector's share of national income. Although output of rice, corn, and soybeans more than doubled, the share of agriculture in national income fell from over one-half in 1968 to one-fourth in 1987. Within the agricultural sector, the contribution of rice to total GDP decreased from almost 20 percent to about 8 percent. The shares of both agriculture and rice in national income stopped their declines after 1980 and increased somewhat through 1987. The declining share of agriculture in national income, even as agriculture itself grew rapidly, illustrates how quickly incomes were growing in the nonagricultural sector.

The share of total employment in the agricultural sector, including rice, also declined between the early 1970s and the mid-1980s, although the number of people employed in the agricultural sector increased. Data in Table 2.3 indicate that virtually all of this decline in the labor share of agriculture took place in the 1970s, when the nonagricultural economy boomed. During the 1980s, the agricultural sector maintained its share of employment at 55 percent because nonagricultural growth fell off with the decline in oil prices and increased agricultural activity took up some of the slack.

Rice Price Level Policy

The immediate concern of the New Order government in the late 1960s was to control consumer prices for rice because the short-term legitimacy of the government was highly dependent on the constituency of urban consumers. Stable and low rice prices were a critical part of the government's economy-wide stabilization program in the first five-year develop-

TABLE 2.3. *Number of people employed in Indonesia by sector, 1971–86*

Sector	1971		1980		1982		1985		1986	
	Million	Percent	Million	Percent	Million	Percent	Million	Percent	Million	Percent
Agriculture	26.47	64.1	28.04	54.8	31.59	54.7	34.14	4.6	37.6	55.1
Mining	0.09	0.2	0.37	0.7	0.39	0.7	0.42	0.7	NA	NA
Manufacturing	2.68	6.5	4.36	8.5	6.02	10.4	5.80	9.3	5.6	8.2
Public utilities	0.04	0.1	0.08	0.2	0.06	0.1	0.07	0.1	NA	NA
Construction	0.68	1.7	1.57	3.1	2.15	3.7	2.10	3.4	NA	NA
Trade	4.26	10.3	6.61	12.9	8.55	14.8	9.35	15.0	NA	NA
Transportation	0.95	2.3	1.47	2.9	1.79	3.1	1.96	3.1	NA	NA
Services	4.21	10.2	7.97	15.6	7.24	12.5	8.57	13.7	NA	NA
Other	1.88	4.5	0.71	1.4	–	–	0.07	0.1	25.1[a]	36.7[a]
TOTAL	41.26	100.0	51.18	100.0	57.79	100.0	62.48	100.0	68.3	100.0

SOURCE: BPS, *Statistical Yearbook of Indonesia*, 1975, 1982, 1985, 1986.
NA = Not available.
[a]For 1986 only, the other category includes all sectors other than agriculture and manufacturing.

ment plan (REPELITA I, 1969–73). A ceiling price for rice in urban areas was announced and enforced on Java as well as on the outer islands. During much of the 1970s, producer and consumer prices were held below world market prices. From 1969 to 1976, a period encompassing first very low and then extraordinarily high world rice prices, the domestic price averaged 30 percent below the border (c.i.f.) price.

The cost of protecting consumers from fluctuations in the world market by subsidizing imports rose considerably with the unprecedented increases in the world rice price during 1973 and 1974. Although government revenues from oil exports grew concurrently because of the quadrupling of the world price of petroleum, for a brief time Indonesia was unable to purchase rice imports at any price. The country's vulnerability to the highly unstable world market conditions at the time gave strong support to the belief that complete self-sufficiency was a desirable strategy to guarantee food security.

During the mid-1970s, the government began to devote more attention to rice production. Indonesia was routinely the largest importer in the world markets during the second half of the 1970s, often accounting for 20 percent or more of world trade. The country's dependence on imports was aggravated by production shortfalls stemming from the widespread infestation of the brown planthopper from 1976 to 1978.

As Indonesia absorbed increasingly larger shares of the world market, policymakers reassessed the desirability of subsidizing rice imports, which was necessary because domestic wholesale rice prices were held by policy at levels lower than world prices. This policy of setting low domestic rice prices was gradually superseded by the government's desire to promote growth of rice output for long-run food security. Relative to world prices, domestic prices were increased steadily during the last half of the 1970s. When world rice prices fell in the 1980s (see Chapter 3), Indonesian price levels consistently were held above the low world prices. After the world price recovered in 1987 and early 1988, prices in Indonesia were raised too. Several influences undoubtedly entered into the government's decision to raise average domestic rice prices relative to the world price: many rice consumers were less in need of subsidized rice prices following two decades of high income growth; a package of Green Revolution innovations (detailed below) was available and could be spread more rapidly under a positive price environment; and the government had adequate budgetary resources to support agricultural development.

Rice Price Stabilization Policy

A domestic buffer stock, intended to achieve stable rice prices and thereby enhance food security, was introduced in 1974 and has been implemented successfully by the National Food Logistics Agency

(BULOG) since that time (Falcon et al. 1985). BULOG defends a floor price to farmers by offering to buy paddy at the village cooperative (KUD) level at the announced floor price, storing purchased grain in government warehouses, and selling rice from stocks when the wholesale price approaches the desired ceiling level. The band between the floor price and the urban retail price is kept large enough to allow for active private participation in the storage and distribution of rice. Reducing the marketing margin to meet the conflicting short-run objectives of low urban prices and adequate incentives to farmers would result in undesirably high storage and administrative costs for the government. BULOG generally has been successful in maintaining a sufficiently wide band to encourage private trade; the agency's procurement of rice has never exceeded 12 percent of total production, and its distribution has been limited to no more than 15 percent of total consumption.

A shift in the ceiling price policy in the early 1980s helped to lower administrative and storage costs for BULOG. A single urban ceiling price is no longer officially announced, and the government allows retail prices in the outer islands (in areas without surplus rice production) to rise somewhat above retail prices on Java. The retail price differential between regions has provided greater incentives for private traders to purchase and distribute rice from surplus to deficit regions.

The wholesale price has rarely fallen beneath the announced floor price, with a notable exception in 1985. BULOG has held enough stocks to last three to four months and has varied its purchases of imports (BULOG has an import monopoly) to keep the buffer stock at desired levels. Trade and public stock data are shown in Table 2.4. Generally, BULOG has been able to defend the official floor price successfully.

Fertilizer Subsidy Policy

Much of Indonesia's success in expanding its rice production is attributable to a combination of output and input price policies that improved the profitability of rice cultivation. Since 1968, the prices of virtually all inputs in rice production, except land, animal power, and farm labor, have been influenced directly by government policy. The costs of seeds, water, fertilizer, pesticides, fuel, and machinery have been reduced at various times by specific price or credit subsidies. The subsidy on fertilizer, in particular, has been a key policy tool in Indonesian rice price policy. Timmer (1985) estimates that as much as one-half of the growth in rice production from 1968 to 1984 was attributed to improved incentives to farmers created by the fertilizer subsidy and stable rice prices.

The wholesale price of rice and the paddy support price have not risen significantly in real terms during the past twenty years (deflated either by

TABLE 2.4. *International trade and BULOG stocks of rice, 1969–88 (tons)*

Year	Rice exports	Rice imports	BULOG closing stock, March 31[a]	BULOG closing stock, July 31[a]
1969		603,200		
1970		955,629	236,134	315,863
1971		493,482	397,461	341,984
1972		733,511	387,477	380,952
1973		1,656,680	198,487	206,756
1974		1,070,770	579,961	931,795
1975		672,665	778,005	1,047,046
1976		1,280,580	521,523	778,169
1977		1,964,070	572,027	827,136
1978		1,838,260	470,252	1,314,227
1979		1,929,490	707,851	1,035,523
1980		2,026,550	885,665	2,253,652
1981		525,442	1,191,950	2,384,759
1982		300,188	1,593,212	2,749,819
1983		1,154,930	911,062	1,549,973
1984	10,979	375,158	1,441,653	2,925,476
1985	406,013	0	2,315,541	3,373,624
1986	231,425	0	2,130,963	2,144,411
1987[b]	118,642	119,950	1,820,525	2,493,069
1988	20,000	21,000	769,201	1,383,117

SOURCE: BULOG.
[a]Figures for 1970–73 are for milled rice only and underestimate total end-month stocks for these years because they exclude unmilled rice in storage.
[b]Figures for1987 include commodity export loans and import repayments of rice to or from the Philippines and Vietnam.

the consumer price index or the GDP deflator). But this constant real rice price gave strong encouragement to increase production because new technology caused real production costs to decline. Input subsidies reduced the costs of the new technology even further. When the paddy support price is measured relative to the subsidized domestic fertilizer price, the change over time has been considerable (see Table 2.1). Since 1968, the government has set the price of urea at the wholesale level and has allowed private traders as well as village cooperatives to distribute urea to farmers. Quantities available for distribution are not restricted. To ensure adequate supplies, large public investments in domestic manufacturing plants for fertilizer were made in the mid-1970s.

Table 2.1 shows that significant incentives to farmers began after 1977, when the ratio of paddy to urea prices was held consistently above 1.0, rising to almost 2.0 in 1982. Nominal prices for urea were held constant at Rp 70 per kilogram from 1976 to 1982 and were raised to Rp 90 per kilogram in 1983 and 1984. Since 1985, the urea price has been increased (to Rp 100 per kilogram in 1985, Rp 125 in 1986 and 1987, Rp 135 in 1988, Rp 165 in 1989, and Rp 185 in 1990) in order to reduce the size of the budget subsidy. Officially reported expenditures on the fertilizer subsidy as a proportion of total development expenditures peaked at over 7 percent in 1984–85, when the fertilizer subsidy was Rp 732 billion. The continuing need for these subsidies is a matter of considerable controversy, both analytically and in policy discussions.

Public Investment Policy in Rice Production

The near doubling of average rice yields between 1968 and 1989, shown in Table 2.1, resulted mostly from the widespread adoption of high-yielding varieties that were developed jointly by researchers at the International Rice Research Institute (IRRI) and at Indonesian rice research stations. HYVs were first released in 1967 and disseminated throughout Indonesia as part of the Ministry of Agriculture's BIMAS (mass guidance) research and extension program. The technology package supporting the adoption of HYVs included input recommendations, subsidized credit, and the availability of subsidized fertilizer and pesticides. The BIMAS program was an important ingredient of rice development policy in the 1970s but declined in the 1980s after most farmers adopted HYVs and were capable of funding inputs from rice profits.

The initial adoption rate of HYVs was rapid and reached one-third of irrigated area by 1974. Subsequent infestations of the brown planthopper decreased the rate of adoption in the mid-1970s, but the development of a new seed variety resistant to the predominant strain of the brown planthopper (biotype 2) raised the HYV adoption rate again after 1977.

Pesticide use by rice farmers increased rapidly, but with very mixed results. By the mid-1980s, farmers paid only 10 to 20 percent of the world price for the most widely used brands. The heavy, subsidy-induced use of several pesticides, especially Diazinon, however, killed the natural predators of planthoppers. The hazards to human health of pesticides were also becoming clearer by the mid-1980s, as were certain financial irregularities within the pesticide program itself. As a consequence, the pesticide subsidy was removed in 1988, and major efforts were made to install a system of integrated pest management. The changed focus received positive reactions throughout the world. The integrated system is new and incompletely tested, however, and pests likely will continue to be an intermittent deterrent to rice production in the 1990s.

The development and dissemination of new seed varieties are critical to the success of integrated pest management and the continued expansion of rice output. Since the 1950s, thousands of local varieties of rice have been replaced by a few rice varieties that are closely related genetically. In the mid-1980s, two rice varieties (IR 36 and Cisadane) accounted for over one-half of the seeds planted in Indonesia's wetlands. The relatively narrow genetic base embodied in these varieties significantly increases their susceptibility to widespread pest infestation.

Varietal research continues at IRRI and in Indonesia. Expenditures on agricultural research in Indonesia have been increased gradually during the 1980s from 3 to nearly 4 percent of agricultural GDP as external aid donors have replaced the declining levels of spending by the Indonesian government (Nestel 1987). Even with increased expenditure, however, the future for new varietal breakthroughs is sobering. There has not been a fundamental breakthrough in experimental yields for twenty years, and the variety IR 8 still sets the yield standard throughout Asia. Progress continues to be made via improved yield stability from resistance to pests and drought, but Indonesia is already on the research frontier with respect to germplasm for many of its irrigated farming systems. Unless there are unexpected contributions from genetic engineering, yield increases during the 1990s will probably be much more difficult to attain than those in the 1970s and 1980s.

The dissemination of advanced rice technologies has been facilitated by investments in marketing infrastructure (roads and ports) and irrigation systems, especially on Java. Investments in irrigation have been particularly critical to achieving success in the adoption of HYVs because the new seed varieties were specifically adapted for irrigated systems. Data indicating the amounts of new and rehabilitated irrigated areas resulting from public investment are presented in Table 2.5. Lowland areas with existing irrigation systems, including many regions on Java, were favored by the

TABLE 2.5. *Additions to irrigated area from public investments, 1969–89 (000 hectares)*

Years	Rehabilitated	New
1969–70	210	43
1970–71	172	24
1971–72	136	46
1972–73	155	46
1973–74	163	31
1974–75	109	21
1975–76	105	89
1976–77	117	63
1977–78	112	41
1978–79	85	112
1979–80	95	123
1980–81	112	113
1981–82	94	118
1982–83	69	57
1983–84	24	26
1984–85	87	69
1985–86	30	61
1986–87	37	22
1987–88	153	60
1988–89	404	24

SOURCE: Data from Varley (1989).

TABLE 2.6. *Agriculture's share of development expenditures, 1974–87*

Years	Total development expenditures (billion rupiahs in constant 1977–78 prices)[a]	Share of agriculture in total (percent)[b]
1974–75	1,558.7	35.5
1975–76	1,952.3	23.8
1976–77	2,322.2	19.5
1977–78	2,157.6	16.5
1978–79	2,411.9	20.1
1979–80	2,867.3	13.1
1980–81	3,792.4	15.7
1981–82	4,180.7	15.5
1982–83	4,066.1	15.3
1983–84	5,050.6	12.6
1984–85	4,544.2	18.8
1985–86	4,727.4	13.0
1986–87	3,443.0	14.4

SOURCES: World Bank, "Policies for Growth with Lower Oil Prices" (May 12, 1983), pp. 196–200; BPS, *Financial Statistics* (various years).
[a]Deflated by the Jakarta Consumer Price Index.
[b]Includes irrigation, excludes project aids.

initial investments in irrigation between 1968 and 1975. That pattern was reversed in later years, when irrigation investment shifted to less productive rice regions, mainly in the outer islands.

The additions to irrigated area, shown in Table 2.5, indicate a marked decline in investment in irrigation and expansion in area during the early 1980s. Additions to irrigated area, including both rehabilitated and new areas, were only about half as large in the mid-1980s as they were in the late 1970s and early 1980s. This pattern of reduced public spending on irrigation resulted from a fall in the total development budget and a decline in the share of that budget devoted to agricultural investment. As shown in Table 2.6, the total real (deflated) development budget reached a peak in 1983–84 and fell to two-thirds of that level three years later, after the oil price decline. Meanwhile, the share of the development budget spent on agriculture (including irrigation) dropped to 12 to 15 percent in the mid-1980s after being double that portion in the mid-1970s. This investment pattern was in the process of being reversed in the late 1980s. Such a reversal, both in total agricultural investment and attention given to improved maintenance of irrigation systems, could have a significant impact on the government's ability to carry out its future rice strategy.

Macroeconomic Policies Affecting Rice Production

Macroeconomic policies, defined in Chapter 1, have had little impact on rice production. Although macro policy cannot provide specific assistance to agricultural production, it could offer positive or negative incentives (even if they were unintended). For example, an overvalued exchange rate would tax most agricultural producers implicitly by causing them to receive too few rupiahs when they sell their commodities (as exports or import substitutes) because the exchange rate is maintained by government policy at an inappropriately low level (i.e., too few rupiahs per unit of foreign exchange).

The history of Indonesian exchange rate policy in the 1970s and early 1980s generally is impressive. Periods of overvaluation, caused mainly by insufficient depreciation of the rupiah (i.e., not enough fully to offset faster Indonesian inflation relative to the inflation rates in major trading partner countries), were followed by large devaluations in November 1978, March 1983, and September 1986. A major reason for the two most recent devaluations was the decline in Indonesia's export prices, especially of petroleum. Although there were periods in which agriculture was taxed by overvalued exchange rates, neither exchange rate policy nor other macro policies had a consistently large detrimental impact on agricultural incentives.

Even when the exchange rate was overvalued, the impact on rice production was negligible. As explained earlier in this chapter, the prices of rice and fertilizers are set by government policy and supported by quantitative restrictions on imports or exports. Consequently, rice farmers are insulated against any taxing effects of overvalued exchange rates because the prices of their output and main tradable input (fertilizer) are determined by commodity policies irrespective of world prices and the exchange rate. For these reasons, macro policy is not an important element of rice policy in Indonesia. Of course, budgetary stringency resulting from the oil price decline has a very significant indirect impact on the rice sector by limiting the availability of public investment funds for agriculture. This important constraint is considered above as part of investment policy.

Crop Regulatory Policy

Crop regulatory policy consists of programs or regulations that control the area farmers are permitted to plant to particular crops (Tabor et al. 1988). It also includes a set of less codified pressures that the government is able to bring to bear on farmers through a system of decentralized governance that reaches from the capital to the most remote villages. For example, the hectarage in rice is reduced relative to what it would be if farmers were free to exercise complete freedom in selecting their crop mixes and rotations because government programs force production of sugarcane and tobacco on farms that otherwise could grow more profitable rice. These area regulations are discussed in Chapters 4 and 7 in the context of analyzing crop substitutions and policies that limit their applicability.

To ensure that targeted areas are planted, the government selects villages for inclusion in the programs on a rotating basis. Rice farmers in the selected villages are then forced to devote one-third or one-half of their irrigated land to sugarcane or tobacco. These programs, which rarely occur together in the same village, are most developed in East and Central Java. Changes in these regulations could lead to significant increases in rice output, as the analysis of alternative rice strategies in Chapter 7 indicates.

Conclusion

Indonesia has experienced a successful Green Revolution in rice production in large part because the government blended price level, price stabilization, and public investment policies to provide positive incentives for expanded rice output. Macroeconomic policy has not been a major negative or positive element because fixed prices for rice and fertilizer have largely insulated rice farmers from exchange rate policy. Crop regulatory policy has had an inimical influence on rice production by restricting the ability of some farmers to plant as much rice as they would like.

Although increases in rice output have been impressive, maintaining a balance between rice production and consumption remains at the forefront of the policy agenda. What policy options are now available to the government, given that domestic rice prices are aligned with world prices and investment funds for rice are limited by the oil-induced budgetary stringency? Would policies that continue to attract resources into rice production lead to significant increases in rice output, efficient income growth, and rural-based employment? Are there important trade-offs among the objectives of food security, efficiency, and income distribution within the available rice strategies? These issues are at the heart of the empirical investigations reported in the second and third parts of this study.

3. The International Rice Market

Eric Monke and Scott Pearson

This chapter explains why the level and instability of the world price of rice are both critical in shaping Indonesian rice strategy and difficult to forecast. For Indonesia, the level of the world price directly affects the desirability of expanding rice output, the budgetary consequences of rice trading when policy decisions cause domestic and world prices to differ, and the income distributional effects of price policy for rice. The instability of the world price influences the need for stabilization measures and the costs and benefits of BULOG's buffer stock program to stabilize domestic prices. The first section of the chapter briefly develops these arguments.

Subsequent sections review recent patterns of international rice trade and prices, focusing on the forces that determine world price level and variability. The small size of the world rice market contributes to a high degree of price variability because minor changes in the balance between production and consumption in individual countries translate into relatively large changes in world market availabilities or demands. The tendency toward instability is accentuated by government policies that influence the rice market. In a policy-dominated trading environment, variation in international trade is used to maintain domestic price targets. Most governments place great importance on holding domestic rice prices stable, and their changing positions in the world market tend to aggravate international price instability.

Indonesian planners and policymakers make many crucial decisions that require judgments about future patterns of world rice prices. But prediction of the world price is a hazardous exercise. The thinness of the world rice market means that world rice prices are difficult to forecast in the short run; small errors in forecasting production or consumption for individual countries become large mistakes in world market forecasts. To forecast long-run rice prices, market analysts have to contemplate future

We gained valuable insights for this chapter from discussions at a two-day seminar on the world rice market, held in Jakarta, Indonesia, during January 1987. Participants from outside of Indonesia included Donald Mitchell (World Bank), Robert Schwartz (Harvard Business School), Ammar Siamwalla (Thailand Development Research Institute), C. Peter Timmer (Harvard University), and the Stanford Project team.

shifts in policies as well as changes in the behavior of consumers and producers. The final section of the chapter reviews plausible scenarios for the future.

World Prices and Indonesian Food Policy

The insulation of domestic prices from highly unstable world prices is one of the principal goals of Indonesia's food policy. Information about world price instability aids policymakers in selecting the appropriate size of BULOG buffer stocks, necessary to achieve varying degrees of food security. The costs of maintaining these stocks need to be compared with the probability of high world prices during years of low domestic production because an alternative to buffer stocks is to increase imports and sell at subsidized prices. Variabilities in Indonesian production and world prices have been closely linked; Indonesia's rice production is more than twice as large as the amount of trading in the world market. Unexpected production shortfalls sometimes cause Indonesia to import, whereas unusually good harvests can result in occasional exports.

Price instability also has a large impact on the government budget. Budgetary consequences occur if the rice price in Indonesia departs from the world price. For example, Indonesia's domestic prices for rice were well above comparable world prices between 1982 and 1986; as a result, in 1985 and 1986 Indonesia was forced to pay very large export subsidies (about $150 per metric ton) to move surplus rice onto the world market. When domestic prices are below world prices, as in much of the 1970s, policymakers must worry about the costs of subsidizing imports. Given the desire for domestic price stability, the only way to avoid budget outlays is to remain self-sufficient—choosing rice and input prices that exactly balance the growth in production with that in consumption. The vagaries of climate, biology, and economic incentives outside of rice price policy make such precision impossible in most years.

The long-run world price also plays a critical role in investment and price policy. Long-run average world prices of rice are efficiency measures in evaluating investment projects (e.g., in irrigation, varietal research, and extension) or in policy analysis of input subsidies (e.g., on fertilizer, insecticides, and pesticides) that lead to increased rice production. The long-run average world price measures the opportunity cost of the planned increase in production because this rice will be either exported or substituted for imports of rice. If, for example, the world rice price were to approach the temporary peak it reached in 1981 (over $400 per metric ton for Indonesian qualities), Indonesia could expand rice production in most parts of the country and export the new output efficiently.

Indonesia's potential influence on world prices complicates the choice of an appropriate efficiency measure. The c.i.f. price acts as the efficiency cost of imports, while the f.o.b. price measures the efficiency value of exports. Normally, these prices differ only by the costs of international transportation and handling and are relatively small. But for Indonesia, c.i.f. and f.o.b. prices also are influenced by the magnitude of Indonesia's participation in the world market. If Indonesia is a large importer, world market prices are forced upward; if Indonesia exports, world market prices move downward. As a result, efficiency prices for the evaluation of import substitution (the c.i.f. price) and exportation (the f.o.b. price) will differ by more than just transportation and handling costs.

These differences will be larger in the short run than in the long run because of unused production capacity in several of the surplus-producing countries. Indonesia's effect on world prices thus will be more important for price variability than for price level. For example, if Indonesia were to import large quantities of rice, prices in the world market would increase immediately. After a short adjustment period, however, world exports would rise, bringing prices back down. During the last fifteen years, for instance, the internationally traded volume of rice has increased by about 50 percent, while world price levels have steadily fallen. The United States, Thailand, and Japan are examples of countries that are capable of increasing exports in response to increased world market demand in the future.

The long-run average world price also serves as the arbiter of the income distributional effects of domestic rice price policy. When domestic prices depart on average from world rice prices, income transfers occur among producers, consumers, and the government treasury. During the 1970s, the domestic price in Indonesia was usually less than the world price; as a result, Indonesian consumers gained, producers lost, and the treasury (through the food agency) had to subsidize rice imports. In the first half of the 1980s, the domestic price exceeded the world price; consequently, policy caused consumers to pay more, while producers received higher incomes. The government collected import taxes through 1984 but paid export subsidies in 1985 and 1986 after self-sufficiency was achieved. Following the rapid rise in world prices during 1987–89, domestic and world prices were at similar levels. But if world prices return to their low levels of 1985 and 1986, expansion of production cannot occur without subsidies from consumers, the government budget, or both.

Recent Changes in World Trade and Prices

Rice import and export statistics for the 1969–86 period are reported in Table 3.1. During the 1980s, world trade averaged about 12 million met-

TABLE 3.1. *Rice trade statistics, 1969–88*

Country/region	Quantity (million metric tons)					Share of world trade (percent)				
	1969–71	1974–76	1979–81	1984–86	1988	1969–71	1974–76	1979–81	1984–86	1988
World trade	8.9	8.8	12.6	11.8	10.6					
Exports										
Burma	0.7	0.3	0.6	0.6	0.1	7	4	5	5	1
China	2.0	2.4	1.2	1.2	0.9	23	27	9	10	8
Pakistan	0.3	0.6	1.1	1.1	1.0	4	7	9	9	9
Thailand	1.2	1.3	2.9	4.4	4.8	14	15	23	37	45
United States	1.7	2.0	2.8	2.1	2.2	19	23	22	18	21
SUBTOTAL	5.9	6.7	8.6	9.4	9.0	67	76	68	80	85
Asia	4.8	5.3	7.8	8.2	7.0	54	60	62	69	66
Imports										
Asia	5.6	4.6	4.6	2.8	NA	63	21	37	24	NA
Middle East	0.5	1.0	1.9	2.5	NA	5	11	15	21	NA
Africa	0.7	0.8	2.2	3.2	2.4	8	9	18	27	NA
Developed market	0.9	1.2	1.7	1.4	1.0	10	13	13	12	NA
Developed CPEs	0.6	0.5	1.2	0.7	0.5	7	6	9	6	NA
SUBTOTAL	8.2	8.0	11.7	10.6	NA	93	92	92	90	NA

SOURCE: United Nations Food and Agriculture Organization, *FAO Trade Yearbook*. The 1988 data are preliminary, reported in the United Nations Food and Agriculture Organization, *Food Outlook Statistical Supplement*, June 1989.
NA = Not available.

ric tons annually, fluctuating between 11 and 13 million metric tons, a substantial increase over trade volumes of a decade earlier, when quantities traded varied between only 8 and 9 million metric tons. But the international market remains tiny in comparison to total production. Trade represents only 3.5 to 4.0 percent of output (in contrast, about 20 percent of wheat production is traded), and the ratio for rice has not varied much in recent years. Between 1974–76 and 1984–86, annual average world production of rice increased by 38 percent (from 343 million [paddy] to 472 million metric tons), a rate of growth comparable to that of the international rice trade.

Five countries—Burma, China, Pakistan, Thailand, and the United States—provided between 70 and 80 percent of total exports of rice during this period. Thailand had the most rapid growth and by 1979–81 had surpassed the United States as the world's largest exporter. In 1984–86 Thai exports exceeded 4 million metric tons, accounting for more than one-third of total export trade. United States exports have been as large as 3.1 million metric tons (in 1981), but magnitudes have depended heavily on government subsidy programs. During the past decade, subsidized exports accounted for about one-third of total sales, primarily through the food aid (Public Law 480) program. Recent Chinese exports are only about half of their level of the early 1970s, while Pakistan's exports increased nearly as much in relative terms. Burmese exports remain stable (except for 1988) in spite of high recent growth in production (a 20 percent increase between 1979–81 and 1984–86).

Asia accounts for over 90 percent of total production and consumption. Asian countries also provide most of the exports; the growth in Thailand's market share is primarily responsible for the increase from 54 to 66 percent in the Asian countries' export share between 1969 and 1988. On the import side of the market, however, Asian countries have become much less important over time. Asian importers have reduced their reliance on international markets, and equally dramatic increases in imports have occurred in countries of the Middle East, Africa, and the developed countries of the Northern Hemisphere.

In the bottom half of Table 3.1 regional aggregations are used to illustrate the structural change in import demand. With the disappearance of Indonesia from the import side of the market, no single country accounts for more than a few percent of total trade. The quantity data show the decline in Asian imports of nearly 3 million metric tons (from 5.6 to 2.8 million). At the same time, total world trade volumes increased by 3 million metric tons; imports by Middle Eastern and African countries each increased by about 2 million, and imports by the developed market economies (primarily Canada and Europe) grew by more than 1 million. By 1985, the Asian import share had declined from 63 to 24 percent, and the Middle East, Africa, and the developed countries (including the developed centrally planned economies) had each become as large.

In Asia, the declines in rice imports mirror production growth in a number of low-income economies, particularly during the last decade. Although Indonesia's growth of rice output is the most impressive in relative terms (a 47 percent increase between 1979 and 1985), large production growth was also achieved in a number of other Asian importing countries, including India (44 percent), Vietnam (43 percent), Sri Lanka (23 percent), and Bangladesh (21 percent). In the Middle East, import growth stems from population increases and, especially, income growth. Iran, Iraq, and Saudi Arabia alone accounted for 1.5 million metric tons of increased imports during 1970–85 (their total imports rose from 0.2 to 1.7 million metric tons). Increased income also spurred import growth in the developed countries. In Africa, poor production performance in both rice and alternative grains and the availability of food aid are major reasons for the increases in imports.

Figure 3.1 provides a graphic description of price movements for the major grains during 1970–89. The international markets for wheat and corn can be considered in concert, primarily because of the wide use of both corn and wheat as animal feed. In most years, one-fifth or more of wheat production is fed to farm animals, and prices for these two grains are tightly linked (Pearson 1990). But price relationships with the rice market are much more volatile. In 1973–75, rice prices reached record highs, primarily as a consequence of consecutive poor harvests in Asia during 1972 and 1973. Although prices of other grains increased sharply as well, rice prices increased more in relative terms. By 1974, the prices of both high-quality rice (represented by the prices of Thai 5 percent brokens) and low-quality rice (represented by the average unit values of Burmese exports) were much higher than those of wheat and corn. These margins remained large for nearly a decade. Siamwalla and Haykin (1983) identify three reasons for the increases in rice prices relative to wheat prices before 1981—slower growth rates of rice production, faster population growth in rice-consuming areas, and higher income elasticities of demand among rice consumers.

After 1981, the production growth of rice was faster than that of wheat and corn, and this reversal provides part of the explanation for the post-1980 decline in relative rice prices. Between 1979 and 1985, world rice production grew by 22 percent (258 to 316 million metric tons of milled rice), world wheat production increased by 19 percent (424 to 502 million metric tons), and coarse grain production grew by only 13 percent (744 to 843 million metric tons) (World Food Institute, 1986). By 1987, Burmese rice prices were only 6 percent higher than wheat prices, the smallest relative premium since 1972.

FIGURE 3.1. *Grain prices, 1970–89*

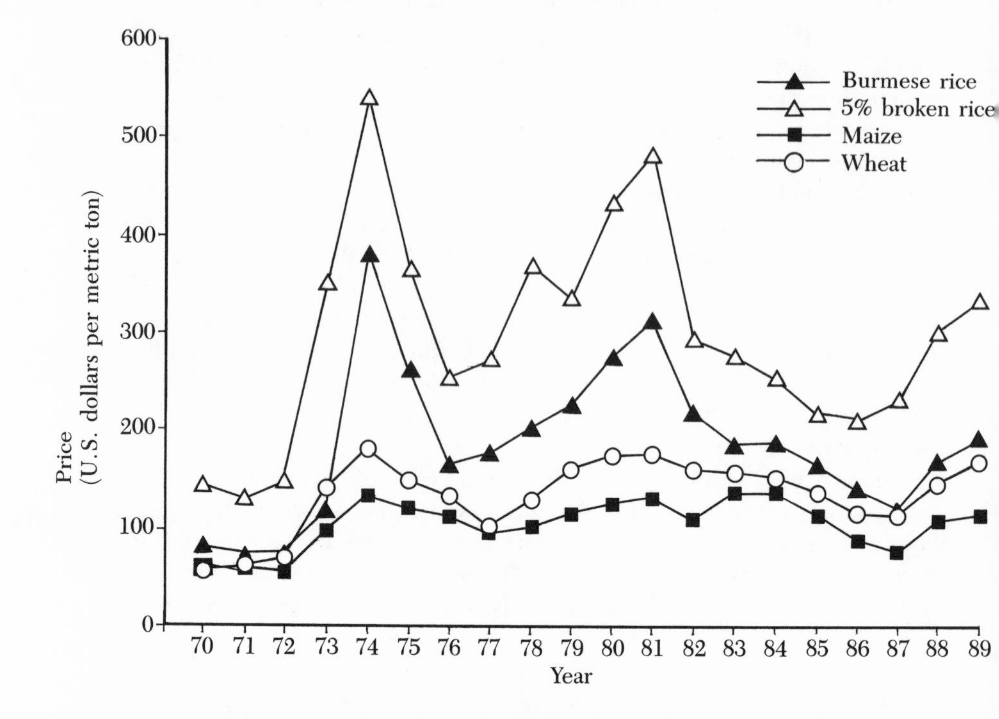

In late 1987, rice prices again increased relative to wheat and corn prices, reflecting the poor weather in Asia. The U.N. Food and Agriculture Organization (FAO) estimates that 1987 production of rice was nearly 4 percent below 1986 production, the first absolute decline since 1972–73. During the spring of 1988, the price of Thai 5 percent brokens rose above $300 per metric ton and fluctuated around that level throughout 1989.

This review of patterns in the main world cereals markets shows that world rice prices are not closely linked to world prices of wheat and corn. Some constraint on the upward movement of rice prices is provided by importing countries substituting wheat and rice. Further, the suitability of rice for animal feed means that rice prices will not fall below corn and wheat prices. The much larger corn and wheat markets, therefore, provide floor prices for the relatively small world rice market. But, in general, the magnitude of price fluctuations in the rice market has been much larger than those in the wheat and corn markets. The world rice price has remained highly sensitive to small changes in the balance between domestic rice production and consumption.

Trade Policies and World Rice Prices

The continued thinness and price variability in the rice market are caused in part by the domestic rice policies. Table 3.2 summarizes information on trade policies used by some of the main participants in the world rice market. Among the exporters, only Australia allows unrestricted trade. The United States and the European Community (EC) countries (primarily Italy) use export subsidy programs; other exporting countries employ either government monopolies or export licenses to set trade limits. Among importers, government monopolies, usually enforced by a national food agency, are widespread. Although the list given in Table 3.2 is not exhaustive, a survey of eighty-six countries, carried out by the FAO in 1983, confirms the prevalence of government monopolies and quantitative controls. These instruments were used by 70 percent of importing countries and 56 percent of exporting countries. A comparison with policies of 1973 shows very little change; although some countries reduced tariff barriers, none eliminated their nontariff controls.

Such widespread and persistent use of quantitative controls and variable taxes or subsidies reflects the effectiveness of trade policy in allowing governments to achieve a wide range of goals—domestic price stability, high prices for consumers or producers, or low prices for consumers or producers. Principal attractions of trade policy are low budgetary costs and ease of administration relative to alternative policies of domestic taxes or subsidies.

Price stability figures prominently in the trade policy decisions of most developing countries. Importing countries increase or decrease international purchases as needed to maintain the desired internal price level; exporters behave similarly with respect to the amounts offered to the international market. Because neither domestic consumers nor producers observe any changes in price, all of the instability must be absorbed in other countries that do respond to world prices. Hence domestic price stability, which is desired in many countries, is achieved only by creating more international price instability and by transferring the needed adjustment to other countries (Falcon and Monke 1979–80; Siamwalla and Haykin 1983; Monke and Taylor 1985). The world price would be more stable if all countries dropped their trade barriers, but under current conditions no country dares to change its policies unilaterally. A stalemate results, and world prices remain highly variable (Monke and Salam 1986).

Low domestic rice prices relative to world prices have been desired in some developing countries such as Indonesia and Thailand in the 1970s. Importing countries can use import subsidies and exporting countries can use export taxes to keep domestic prices below world prices. For many developing country governments, trade policies have been preferred to

TABLE 3.2. *Government and rice trade policy of major participants*

Country	Trade policy
Australia	No controls
Bangladesh	Government monopoly
Brazil	Government licensing (Food Supply Council [CONAB])
Burma	Government monopoly (Myanma Export-Import Corporation)
China	Government monopoly (China National Cereals, Oils, Foodstuffs Import-Export Corporation)
Cuba	Government monopoly (ALIMPORT, Ministry of Foreign Trade)
Egypt	Government monopoly (Rice Mills Organization, Foreign Trade Organization)
European Community	Variable levy and export restitution programs
Hong Kong	Importers are licensed and given quotas determined quarterly by the government
India	Government monopoly (Food Corporation of India)
Indonesia	Government monopoly (BULOG)
Iran	Government monopoly
Iraq	Government monopoly
Japan	Government monopoly (Food Agency)
Korea (South)	Government control (Ministry of Agriculture and Forestry)
Malaysia	Government control (National Padi and Rice Authority). Private importers are licensed, granted quotas, and required to purchase a portion of government-owned domestic rice.
Pakistan	Government monopoly of high-grade basmati rice, government control of lower-grade rice exports through licensing of private traders; export taxes (Trading Corporation of Pakistan Rice Export Corporation)
Philippines	Government monopoly (National Grains Authority)
Sri Lanka	Government monopoly
Thailand	Export permits required for private trade; government uses rice export levies, reserve requirements, and quotas, depending on domestic and world market conditions. Interventions have been little used in 1980s.
United States	Export restitution programs; Commodity Credit Corporation provides export subsidies when world prices fall below support prices plus marketing costs
USSR	Government monopoly
Vietnam	Government monopoly

SOURCES: Falcon and Monke (1979–80), p. 286; Barker, Herdt, and Rose (1985), p. 193; Food and Agriculture Organization, Committee on Commodity Problems, Intergovernmental Group on Rice, *Report* (various years).

direct consumer subsidies to reduce budgetary expenditures. Import subsidies lower market prices and force domestic producers to provide some of the desired subsidy to domestic consumers; export taxes contribute to budget revenues. Because the desire for price stability usually exists concomitantly with the goal of low price levels, variable levies, quantitative controls, and government monopolies on trade often have replaced trade taxes and subsidies as the instruments of intervention.

A high producer price has been prominent in the rice policies of many developed countries—the United States, the EC, and Japan—but also has been evident in developing countries desiring to promote rice production. Indonesian policy in the mid-1980s illustrates this policy in a developing country. The attraction of trade policy relative to direct subsidies to producers is again linked to budgetary implications. In importing countries, tariffs cause prices on the domestic market to rise. Domestic consumers thus provide the desired subsidy to domestic producers, and the government gains revenues from import taxes. In exporting countries, the provision of export subsidies permits the government to maintain a favorable price for farmers without adding to government-owned stocks. Domestic consumers again provide some of the subsidy to producers because they are forced to pay prices that are above those in the world market. The simultaneous desire for domestic price stability mandates the use of quantitative trade controls or variable levies in place of trade taxes.

Domestic policy objectives have almost always taken precedence over international market incentives. Table 3.3 reproduces historical information gathered by Barker, Herdt, and Rose. This table shows the consistency of internal price differences among three groups of Asian rice economies between 1961 and 1980. The highest domestic prices were in high- or middle-income countries—Japan, South Korea, Malaysia, and Taiwan. Domestic prices were near the world price in China and in the low-income, traditional importing countries (some of which are now self-sufficient)—Indonesia, Bangladesh, Philippines, India, and Sri Lanka. Very low domestic prices arose from taxing policies in three exporting countries—Thailand, Pakistan, and Burma. In some of the countries, the pattern has changed over time. Indonesian producer prices have risen substantially in the last decade relative to world prices; Sri Lankan prices have fallen even more dramatically. These changes correspond to fundamental shifts in domestic policies: in Indonesia, the achievement of self-sufficiency and agricultural income growth; in Sri Lanka, the emergence of government and consumer budget constraints on high producer prices.

TABLE 3.3. *Farm-level price of rice as a percentage of world price, 1961–80*

Country	1961–65	1966–70	1971–75	1976–80
Japan	203	228	246	391[a]
South Korea	119	104	111	187
West Malaysia	–	–	149[b]	173
Taiwan	160	134	150	168
Indonesia	–	63	66	98
Bangladesh	127	126[c]	163[d]	93
Philippines	120	93	85	77
China[e]	109[f]	96	71	76
India	146	109	98	76
Sri Lanka	178	141	128	76[g]
Thailand	71	55	62	70
Pakistan	–	–	–	48[h]
Burma	56	42[e]	44[e]	37[e]

SOURCE: Barker, Herdt, and Rose (1985), table 16.2, copyright © 1985 by Resources for the Future, Washington, D.C.

NOTE: Ranked from high to low based on 1976–80 percentages. Farm-level price in "paddy-equivalent." World price based on quantity and value of total world exports and imports as reported by FAO. World price divided by 2 to obtain paddy equivalent. Paddy-equivalent farm price for each country divided by paddy-equivalent world prices and multiplied by 100 to obtain percentage of world price.

[a]1976–78 only.
[b]1975 only.
[c]1966–67 only.
[d]1972–73 and 1975–76 only.
[e]Official procurement price used.
[f]1965 only.
[g]1976–77 and 1979–80 only.
[h]1977–79 only.

Policies and Future Changes in the World Rice Market

World price effects can arise from any of a large number of possible shocks to the international demand-supply balance in major producing countries. A number of stochastic events, if transmitted to world markets, could cause international prices to rise: disease infestation or pest attacks on high-yielding varieties; drought in any of the major producing countries; a reduction in producer support in developed countries (such as Japan); political instability in major producing or consuming countries causing reduced exports or increased imports; and emergence of a new source of import demand. Although the probability of any one of these events is small, the joint probability is high. When that probability is accompanied by trade policy controls in many of the market participants, world prices can be expected to fluctuate frequently.

The world price of rice peaked four times in the past two decades—in

1974, 1978, 1981, and 1988—and fluctuations are likely to continue at similar frequencies in the future. Problems from weather, diseases, and pests will continue to cause relatively large variations in market demand and supply. The increased emphasis given to accumulations of domestic stock in some countries, such as India and Indonesia, may moderate the magnitude of future fluctuations relative to those of the 1970s. But policymakers' demand for food reserve stocks has proven nearly as inelastic as the demand for imports to meet domestic consumers' needs, and stock policy seems unlikely to reduce the frequency of fluctuations.

The prominence of policy interventions in the world rice market means also that forecasts of long-run world prices are hazardous. Econometric models can confirm the importance of domestic policy objectives in explaining world market performance and can show that world prices are unstable because government institutions transmit variabilities in domestic production to the international market. But such models are not useful for making price forecasts because predictions of future changes in government rice policies are so precarious. Policy changes depend on alterations in political environments and political responses to varying economic conditions. Statistical models are of little assistance in predicting these changes. For example, economists in the mid-1970s were unable to foresee the major changes in Indonesian or Chinese agricultural policies. Indonesia emphasized positive production incentives and reduced rice imports by 2 million metric tons, whereas China encouraged crop diversification and reduced exports of rice by 1 million metric tons. Such changes had direct implications for world prices.

Predictions of long-run rice prices thus depend on judgmental analysis of the interactions between likely economic and policy changes. The marked shifts in import shares in recent years provide some reason to think that international demand may be more stable than in the past. Between 1970 and 1985, the import share of the Middle East and the developed countries doubled—from 22 to 44 percent. The addition of import shares of high-income countries from other regions (such as Hong Kong, Malaysia, and Singapore) means that well over half of world import demand in the mid-1980s arose in countries with relatively high per capita incomes. Most of these countries have little or no local production of rice. Demand for imports tends to be highly price-inelastic because high-income consumers are insensitive to prices, and stable, because income elasticities are near zero and domestic production is unimportant. But increased stability of total demand depends largely on the ability of former importing countries—particularly in Asia—to sustain production growth in the future.

Among rice-producing countries, key policies involve investment and prices. Investments in new technology, including expenditures on research to develop modern varieties and improved farming and processing practices, subsidies on inputs (especially fertilizer), and investments in irrigation and other farm and postfarm infrastructure can have enormous impacts on supply. If such investments result in global expansion of rice output at rates exceeding those of rice consumption, the world rice price level will continue to decline. Technological change in rice (and in other cereals) has been sufficiently rapid in the past to permit the trend of real prices of rice (and of wheat, corn, and other coarse grains) to fall slightly during the past several decades. The interesting question is whether this gradually declining trend is likely to continue.

Barker, Herdt, and Rose (1985) project likely changes in both supply and demand for eight Asian countries (China, India, Indonesia, Bangladesh, Thailand, Burma, Philippines, and Sri Lanka) that together account for 85 percent of Asia's rice production (and thus over three-fourths of global rice output). Table 3.4 shows the results of their base-run projection of production for the year 2000. This projection assumes that irrigated area grows at historic rates, modern varieties spread wherever possible, fertilizer availability increases at a rate of 5 percent per year, and productivity of inputs does not change. The result of this projection, 409 million metric tons of paddy, is much less than their projection of demand by these eight countries for 2000, 482 million metric tons. This demand estimate assumes lower population growth rates than those that occurred in the 1970s but similar rates of growth of income to those of the 1970s and declining income elasticities of demand.

The authors next change a critical policy assumption and project the results of a doubling in the historic rates of growth of irrigated land in seven of the eight countries (excluding China); irrigated area in 2000 then climbs from 54 percent to 62 percent of total area, and projected production expands to 466 million metric tons, approximately the level of the demand projection. Finally, they relax the antihistorical assumption of constant productivity and make projections of the effects of productivity increases for fertilizer, irrigation, and both fertilizer and irrigation combined; the resulting projections of rice output are 460, 499, and 543 million metric tons, respectively.

This exercise shows the difficulty of making price projections for a market in which price and investment policies dominate. Whether growth of the supply of rice in Asia is likely to fall short of, just keep up with, or exceed growth of demand depends in large part on what is assumed for investment policy, especially in irrigation, and for price policy and productivity growth. Projections that run far into the future are particularly vulnerable to errors of being too optimistic or pessimistic about the speed and effects of future technical change.

TABLE 3.4. *Base-run projections of production, consumption, and prices for selected Asian countries for the year 2000*

Country	Production[a] (million metric tons)	Fertilizer (kilograms per hectare)	Percent area		Annual irrigation investment (million $U.S.)
			Modern varieties	Irrigation	
China	196.1	148	65	94	238
India	99.4	67	68	51	576
Indonesia	34.1	89	74	84	457
Bangladesh	28.7	32	63	24	86
Thailand	23.8	25	18	41	82
Burma	14.7	71	56	21	99
Philippines	9.6	61	89	52	37
Sri Lanka	3.1	102	73	66	48
TOTAL OR AVERAGE	408.8	75	64	54	1,623

SOURCE: Barker, Herdt, and Rose (1985), table 18.4, copyright © 1985 by Resources for the Future, Washington, D.C.

[a]Rough rice.

This central point can be illustrated by Indonesia's recent experience with investment and price policies for rice production, detailed in Chapter 2. Indonesia's recent success in expanding output to achieve trend self-sufficiency in rice was not predicted. Appropriate technology policy combined with high incentive price policy to create conditions for rapid growth of rice production, especially between 1977 and 1984. But these policies were complemented and made possible by a stable political situation, macroeconomic policy that did not regularly tax agriculture, petroleum revenues that permitted investments in agricultural and transportation infrastructure, the development of capable individuals and institutions to carry out analysis and to implement policy, and freedom from sustained periods of losses caused by poor weather, disease, or pests. Sensible projections of the future impacts of policies on rice production, consumption, and trade must consider these complementary forces as well.

The future direction of price trends will depend primarily on the conduct of policy in the Asian countries. Other importing regions are small relative to the size of the international market, and production capacity in the major exporters is probably adequate to satisfy increased demands from these regions. Barker, Herdt, and Rose (1985) provide evidence that in the larger Asian countries production capacity is adequate to maintain pace with demand, if policies become more supportive of domestic producers. Such changes probably will be forthcoming. The agricultural sector has become increasingly important in economic development plans as better ways are sought to address problems of employment and income distribution; the recent experiences of Indonesia and India provide successful examples. Although the timing of policy changes is difficult to predict—and such changes could occur partially in response to a rising world rice price—their introduction should be adequate to prevent any substantial increases above the 1989 price levels of $300 to $350 for Thai 5 percent brokens. Prices could be even lower if investment activity greatly accelerates or new technologies are developed. In this event, prices can be expected to continue their long-run decline.

Conclusion

The level of the world price of rice influences Indonesian food policy because it serves as an indicator of the value of additional rice output and sets bounds on the income distributional effects of price policy. The instability of the world price of rice has encouraged the creation of a buffer stock policy, carried out by BULOG, to insulate domestic rice prices. Both the level and instability of world prices have implications for the government budget. Consequently, any Indonesian strategy for rice production, consumption, and trade needs to be grounded in a careful analysis of past and future world rice prices.

World rice prices are influenced heavily by government price policies that insulate producers and consumers from international price instability and by technology policies that increase output and productivity through public investments. Because of the large impact of policies, efforts to project future world prices are hazardous. Econometric models have difficulty incorporating these policy effects and are of little help in predicting them. Less sophisticated exercises to contrast projected growth in demand with that in supply indicate that future growth in rice production has the potential to match growth in consumption.

Indonesian policymakers can expect the world price of rice to continue to be highly variable and difficult to predict. The best guess for the expected future world price trend is that there will be little or no change in the trend of the real rice price. But this outlook—and especially year-to-year world price fluctuations—are influenced heavily by government policies and possible changes in them. In addition, world rice prices in any given year are buffeted about by supply shifts caused by good or bad weather and political disturbances. Perhaps the central lesson for rice policymakers in Indonesia is to measure the impact of their policies relative to the world price trend while implementing flexible trade policies that minimize the costs of domestic price stabilization.

4. Rice Production Systems

Paul Heytens

Rice is Indonesia's most important food crop. Harvested area, just under 10 million hectares per year, is roughly three times the area devoted to corn and eight times that planted to soybeans. Production of rice is almost entirely a smallholder activity and provides income and a staple food for perhaps 20 million households (almost 100 million individuals), which rarely operate more than two hectares of land and often less than one.

Rice is grown in a diversity of agroclimatic zones in Indonesia. In any particular region, individual farmers cultivate and manage rice and competing crops on land representative of several different cropping environments. For this reason, changes in price signals cannot be expected to have the same impact on all farmers in any particular region. For purposes of analyzing alternative rice policies, the wide diversity of Indonesia's rice environments must be averaged into a manageable number of representative rice production systems.

Wetland Rice Production Systems

Farmers were surveyed in five major rice-producing areas to provide a basis for characterizing the important production systems. The five areas (see Figure 4.1)—Kediri in East Java, Klaten in Central Java, Majalengka in West Java, Agam in West Sumatra, and Pinrang-Sidrap in South Sulawesi—are agroclimatically diverse and important rice producers. The survey approach used at each fieldsite consisted of a series of general exploratory interviews conducted in selected villages. On average, about twenty-five villages and seventy-five farmers from a representative sample of agroclimatic zones were visited per site.[1] Tremendous diversity in climate, relative factor endowments, and available infrastructure was ob-

[1]With the exception of sugarcane, the crop budgets are not based on statistical averages of sample farmers. Instead, the Stanford team created synthetic budgets to be representative of the particular crop and system in each of the fieldsites in the farm survey. The use of synthetic budgets reflects the Stanford Project's objective of gathering a broad spectrum of information concerning rice production and labor market conditions in major producing areas during a relatively short period of time.

FIGURE 4.1. *Five major rice-producing areas in Indonesia*

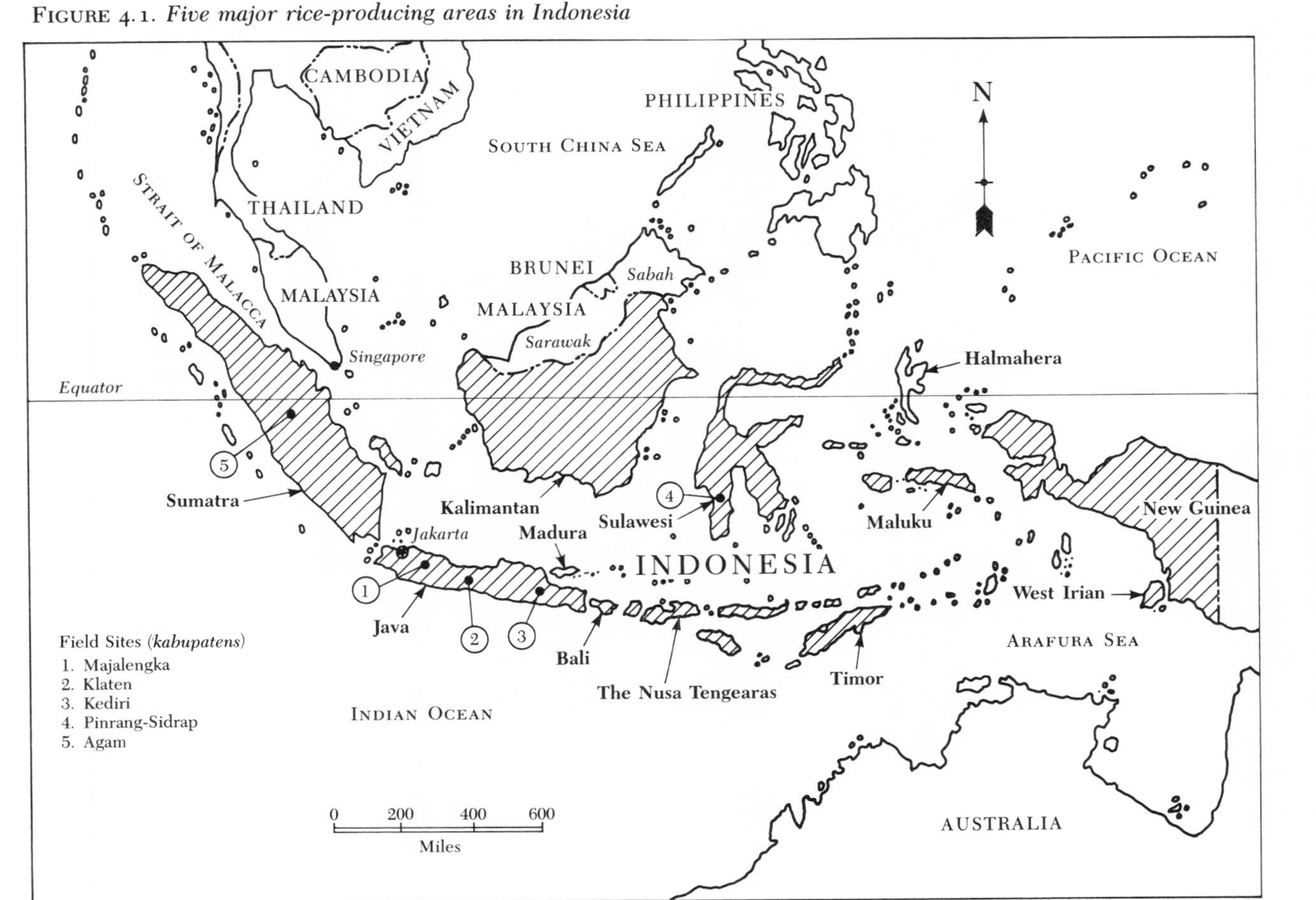

TABLE 4.1. *Wetland rice systems in Indonesia*

	Irrigated *sawah*			
	Good control	Moderate control	Poor control	Rainfed *sawah*
Percentage of				
Total rice area	11	21	26	20
Java rice area	20	30	26	18
Off-Java rice area	2	11	27	21
Yield (tons per hectare)	>6	4.5–6	<4.5	2.5–4.5
Fertilizer use (kilograms per hectare)	400+	400+	250–400	0–300
Crop system[a] (predominant pattern listed first)	rice-rice-rice rice-rice-*palawija*	rice-rice-*palawija*	rice-*palawija-palawija* rice-rice-*palawija* rice-*palawija*	rice-*palawija*

SOURCE: Field surveys, interviews with personnel from BPS.
[a]Wet season crop is listed first.

served. But the survey results can be generalized into the four representative systems found in Table 4.1.

The major land type used for rice production in Indonesia is *sawah* or floodable, bunded cropland. *Sawah* can be divided into irrigated and rainfed typologies. Controlled and timely applications of water, along with effective drainage, are prerequisites for achieving high yields with HYV seeds. Adequacy of water supply and control over its use, therefore, distinguish wetland rice production systems. Three major irrigated systems can be defined largely according to differences in control over water supplies. Water conditions on rainfed *sawah*, although very different from those on irrigated land, are fairly similar throughout Indonesia so only one rainfed system is formally classified. Differences in water conditions, in turn, result in different uses of inputs, cropping intensities, and levels of productivity.

The irrigated and rainfed *sawah* systems illustrated in Table 4.1 accounted for 78 percent of cultivated area and 90 percent of total rice production in 1987 (the corresponding figures were 94 and 97 percent, respectively, for Java). These systems are found predominantly in lowland areas (those below 800 meters). Rice production in Indonesia also occurs in dryland, swamp, and tidal/swamp environments. These environments have not been an important focus of government intensification programs to date and often are found in remote areas poorly serviced by transportation and marketing infrastructure. Consequently, production is characterized by tiny percentages of marketed output, low yields, and modest or zero levels of modern inputs. Table 4.2 provides a rough estimate of the shares of rice area and production contributed by each of the main rice systems.

Each of the systems is well represented throughout Indonesia. Even in seemingly homogeneous regions serviced by large-scale technical irrigation systems, such as the northern coastal area of West Java, variations in the level and timing of water deliveries, drainage, and system management result in the presence of several systems.

TABLE 4.2. *Distribution of Indonesian rice systems*

	Cultivated area (percent)	Total production (percent)
Good-control *sawah*	11	17
Moderate-control *sawah*	21	27
Poor-control *sawah*	26	27
Rainfed *sawah*	20	18
Swamp/tidal	11	6
Dryland	11	5

SOURCE: Estimated from various BPS publications.

The economic and agronomic profiles of the four *sawah* systems, described below, focus on the distinguishing features of crop production. There is considerable diversity even within systems, and descriptions depict only the most typical conditions. The main purpose of the summaries is to highlight the important influences on crop substitution and productivity in wetland rice production. In the rice-producing areas of Indonesia, farmers make basic agronomic and economic decisions for two or three crop seasons per year. As will be shown, the profitability of growing rice and competing crops varies widely among the production systems and seasons.

Budgets from the survey sites for rice and nonrice crops add a quantitative dimension to the analysis. In the budgets, input and output prices are allowed to vary by site and season to reflect local market prices. The appendix at the end of the chapter contains a summary of the values used in computing the budgets as well as a brief explanation of survey methodology.

Good-Control, Irrigated Sawah

The system with the best water control is also Indonesia's most productive. Good-control *sawah* accounts for 11 percent of cultivated area in rice and 17 percent of total annual production. Examples of this system are concentrated in the rich, volcanic soil areas of lowland Java. Off-Java examples are found in the well-irrigated areas of South Sulawesi, Sumatra, and Bali, which together account for less than 20 percent of the total high-productivity area.

Good-control *sawah* is characterized by at least two rice crops per year, often three, and technical or semitechnical irrigation with a reliable water source.[2] Local rainfall has little impact on cropping patterns, except when extreme drought affects reservoir levels and river flows. Further, because of the presence of technical and semitechnical irrigation with permanent controls up to secondary and tertiary canals, drainage problems rarely limit the production of nonrice (*palawija*) crops.

Yields on good-control *sawah* average over 6 tons of dry, unhusked rice (*gabah*) per hectare. Land productivity often is highest in the dry seasons because ample irrigation water enables the rice plant to receive full benefit from the more favorable solar conditions prevailing in dryer periods. Drought stress, therefore, is rarely a problem in the dry seasons. Flood-

[2]Technical irrigation provides maximum control over water flows by using permanent canals, control structures, and measuring devices. The government typically controls water distribution up to the tertiary canals. Semitechnical systems have permanent canals but few controls or measuring devices, and the government generally controls only the source and the main canal. Simple systems have few permanent control and distribution structures and are usually farmer-managed.

ing and lodging can affect yields in the wet season when severe weather occurs.

In this stable and productive crop environment, farmers are able to exploit Green Revolution technologies. Adoption of improved, national variety seeds (e.g., Cisadane) or IR varieties (developed initially at the International Rice Research Institute) is almost universal. Fertilizer use is very high, ranging between 400 and 600 kilograms of urea and triple superphosphate (TSP) per hectare, and ammonium sulfate (ZA) and potassium chloride (KCl) are frequently applied. Both liquid and granular pesticides are used when conditions dictate. In areas where labor costs are high, such as in certain well-irrigated parts of West Java, West Sumatra, and South Sulawesi, mechanization of land preparation using tractors has occurred.

Farmers prefer to plant rice whenever possible (except to break pest cycles or to restore land fertility by fallowing). The figures in Table 4.3 indicate that rice is very profitable on good-control *sawah*, and cash returns ranged from Rp 700,000 to 1.3 million per hectare in 1987, well in excess of prevailing land rents.[3]

Table 4.4 shows that corn and soybeans, rice's major competitors in Java, are much less profitable than rice. This is particularly true in the dry seasons, when returns to rice are highest because of seasonally higher output prices and higher yields. Farmers in good-control areas on Java plant *palawija* crops only for rotational reasons or when rice is not possible because of insufficient water. The crop substitution pattern in off-Java areas is similar, except that *palawija* crops are much lower yielding and thus much less profitable than on Java. Farmers surveyed in Pinrang-Sidrap, South Sulawesi, and Agam, West Sumatra, for example, often left *sawah* fallow rather than plant *palawija* crops when water was insufficient for rice.

In much of Java, villages with good-control *sawah* are often targeted for either the government's sugarcane (TRI) or tobacco intensification programs (but rarely for both at the same time). If a village is chosen for either program, typically 33 percent (sugarcane) to 50 percent (tobacco) of total *sawah* must be set aside for a period ranging from two years for tobacco to several years for sugarcane. Among survey farmers, participation in either program is regarded as a form of forced cultivation.

[3]The level of rice profits seems excessive in comparison with prevailing land rents. During the survey, it appeared that the land market was in disequilibrium at most fieldsites, and survey respondents often had difficulty providing a current land rental rate when asked. *Sawah* land rents have been increasing in most survey villages but at the time of the survey had not caught up with the increases in rice profits, which have risen rapidly and steadily throughout the 1980s (see Chapter 6 for an estimate of the progression of rice profits since 1969).

TABLE 4.3. *Costs and returns for rice on good-control, irrigated* sawah, *1987*

	Klaten, Central Java		Majalengka, West Java		Agam, West Sumatra	
	Wet (Jan–April)	Dry (May–Aug; (Sept–Dec)	Wet (Nov–Feb)	Dry (Mar–Jun; Jul–Oct)	Wet (Nov–Feb)	Dry (May–Aug)
Returns (000 rupiahs per hectare)						
Yields (kilograms per hectare)	7,000	8,000	7,000	7,000	6,500	6,500
Value (rupiahs per kilogram)	200	225	175	200	235	235
Gross value	1,400	1,800	1,225	1,400	1,528	1,528
Value of harvest share	93	120	245	280	153	153
Value of farmer's share	1,307	1,680	980	1,120	1,375	1,375
Costs (000 rupiahs per hectare)						
Cash input costs	76	81	64	72	76	76
Cash labor costs	212	212	188	198	429	429
Cash return[a]	1,019	1,387	728	850	870	870
Implicit family labor	49	60	55	68	60	60
Land rent	225	225	175	175	547	547
Implicit interest cost[b]	12	12	10	11	20	20
Net Return[c]	733	1,090	488	596	243	243

SOURCE: Field surveys.
[a]Value of farmer's share less cash input and labor costs.
[b]1 percent per month for four months.
[c]Cash return less implicit family labor, land rent, and interest cost.

TABLE 4.4. *Costs and returns for* palawija *crops, dry seasons, good/moderate-control, irrigated* sawah, *1987*

	Agam	Kediri, East Java				Klaten	Majalengka
	Soybeans	Arjuna corn	Hybrid corn	Second-season soybeans	Third-season soybeans	Soybeans	Soybeans
Returns (000 rupiahs per hectare)							
Yields (kilograms per hectare)	1,000	4,200	5,000	1,100	1,400	1,300	800
Value (rupiahs per hectare)	600	170	170	600	600	700	700
Gross value	600	714	850	660	840	910	560
Costs (000 rupiahs per hectare)							
Cash input costs	101	141	197	90	152	83	68
Cash labor costs	493	151	163	204	300	222	187
Cash return	6	422	490	366	388	605	305
Implicit family labor	26	15	17	4	4	21	23
Land rent	175	250	250	250	250	200	150
Implicit interest cost[a]	18	12	11	9	14	9	8
Net return	−223	145	212	103	120	375	124

SOURCE: Field surveys.

[a]1 percent per month for three months.

Farmers designated to participate in the tobacco program in *Kabupaten* Klaten, Central Java, for example, rent their land to a tobacco firm and do not cultivate or pay production costs. In 1987, the government offered farmers Rp 785 per kilogram of tobacco harvested from their land. With an average yield of 1.2 tons per hectare, gross cash returns to the farmer were Rp 942,000 per hectare. The opportunity cost of tobacco (grown from June until December) is one or two dry-season rice crops. Cash returns from only one dry-season rice crop in 1987 were about Rp 400,000 per hectare higher than those from tobacco. Tobacco farmers have time to engage in other activities, but even if farmers were able to find agricultural work in lieu of cultivating rice, earned income from such activities over one rice season is unlikely to exceed Rp 300,000 in Klaten. Off-farm work thus does not compensate fully the opportunity cost of not cultivating rice.

Participants in the sugarcane program pay production costs, but all management and cultivation is handled by a group leader, who receives a percentage of gross revenues. As Table 4.5 indicates, cash returns to sugarcane cultivation on *sawah* in the three Java fieldsites ranged from Rp 522,000 per hectare in Majalengka, West Java to Rp 1 million per hectare in Kediri, East Java. The opportunity cost of sugarcane on *sawah* is three to five crops (at least two of which can be rice), depending on the farmers' ability to squeeze in rice crops before and after sugarcane cultivation. In a scenario of three lost rice crops, sugarcane harvested in August 1987 excluded a third-season rice crop in 1986 and wet- and second-season rice crops in 1987. Cash returns in the three Java fieldsites for these rice crops ranged from Rp 2 to 3.5 million per hectare, a much higher return than that obtained from sugarcane cultivation. Even if farmers could find other

TABLE 4.5. *Costs and returns for sugarcane, good/moderate-control* sawah, 1987

	Klaten	Majalengka	Kediri
Returns (000 rupiahs per kilogram)			
Processed sugar yield (kilograms per hectare)	8,000	8,000	8,500
Output value (000 rupiahs per hectare)	0.4675	0.4675	0.4675
Gross value	3,740	3,740	3,974
Value of farmer's share	2,319	2,319	2,464
Costs (000 rupiahs per hectare)			
Cash input costs	474	544	441
Cash labor costs	717	992	707
Management fee	34	39	79
Explicit interest cost	138	222	173
Cash return to farmer	956	522	1,064
Land rent	675	525	750
Net return to farmer	281	−3	314

SOURCE: Field surveys.

employment as agricultural laborers, rice farming remains by far the more lucrative alternative.

Moderate-Control, Irrigated Sawah

Moderate-control *sawah* is also concentrated in Java, although not so heavily as good-control *sawah*. This second *sawah* system accounts for 21 percent of total cultivated rice area and 27 percent of Indonesian rice production. The features of this system are similar to those of good-control *sawah*, but yields are lower and rice-cropping intensity is not so high because soil and water conditions are less favorable. Yields range between 4.5 and 6 tons of unhusked rice per hectare and typically fall in the dry seasons because of moderate drought stress and pests. Two rice crops per year are almost always feasible, and often, particularly on Java, a third, nonrice crop is grown. Three rice crops generally are not possible because of a lack of water in the third season. Drainage problems prevent cultivation of *palawija* crops during the wet season.

The use of improved seeds is almost universal, and fertilizer application is high, ranging between 350 and 500 kilograms total per hectare. Pesticide use is also high. Like its good-control counterpart, moderate-control *sawah* is located mainly in lowland areas, and mechanization of land preparation is common in regions where wages are high or land preparation needs to be done quickly.

Survey results revealed that rice grown on moderate-control *sawah* is also very profitable. Cash returns in the survey sites ranged from Rp 500,000 to 850,000 per hectare. Returns for rice were slightly higher in the dry season because of higher output prices and are much higher than those for competing crops. This is true even in Kediri, arguably the most profitable *palawija* production area in Indonesia. A comparison of Tables 4.6 and 4.4 shows that rice maintained a comfortable profitability margin over both corn and soybeans in that East Java *kabupaten*. Yields or prices of hybrid corn, the closest competitor, would have to rise roughly 20 percent to match the net profitability of dry-season rice in Kediri. Comparison of Tables 4.5 and 4.6 indicates that rice grown in moderate-control *sawah* in Kediri or Klaten is much more profitable for farmers than is sugarcane.

In Pinrang-Sidrap and Agam, experience with *palawija* crops is more limited. But there, too, rice is considerably more profitable than the nonrice alternatives. Many farmers in South Sulawesi stated that, because of low returns, they would rather leave their land fallow than grow *palawija*. Nonrice crops were more in evidence in Kabupaten Agam, West Sumatra, but very low yields caused net soybean returns there to be negative. Because nonrice crops are less profitable in all areas, they tend to be grown in the dry seasons only when water is insufficient for rice.

TABLE 4.6. *Costs and returns for rice, moderate-control, irrigated* sawah, 1987

	Klaten		Kediri			Pirang-Sidrap, South Sulawesi		Agam	
	Wet (Dec–Mar)	Dry (Apr–Jul)	Wet (Nov–Feb)	Dry 1 (Mar–Jun)	Dry 2 (Jul–Oct)	Wet (Oct–Jun)	Dry (Mar–Jun)	Wet (Mar–Jun)	Dry (May–Aug)
Returns (000 rupiahs per hectare)									
Yields (kilograms per hectare)	5,500	5,000	5,750	5,250	4,750	6,000	5,000	5,000	5,000
Value (000 rupiahs per kilogram)	0.200	0.225	0.180	0.210	0.240	0.160	0.225	0.235	0.235
Gross value	1,100	1,125	1,035	1,103	1,140	960	1,125	1,175	1,175
Value of harvest share	73	75	104	110	114	96	113	118	118
Value of farmer's share	1,027	1,050	931	993	1,026	864	1,012	1,057	1,057
Costs (000 rupiahs per hectare)									
Cash input costs	77	83	83	90	93	78	81	69	69
Cash labor costs	221	221	153	166	166	85	85	429	429
Cash return	729	746	695	737	767	701	846	559	559
Implicit family labor	55	66	26	37	52	53	61	58	58
Land rent	175	175	250	250	250	393	465	415	415
Implicit interest cost	12	12	10	10	10	7	7	20	20
Net return	487	493	409	440	455	248	313	66	66

SOURCE: Field surveys.

Poor-Control, Irrigated Sawah

Poor-control *sawah* is Indonesia's largest and most diverse rice system, accounting for 26 percent of cultivated rice area and 27 percent of production. This system is about equally divided between Java and the outer islands. Unlike the systems where water availability and delivery is highly controlled, poor-control, irrigated *sawah* is found at elevations above 800 meters as well as in lowland areas. This system includes much of the picturesque terraced hillside rice paddies for which Indonesia is famous.

Poor-control *sawah* is characterized by one or two rice crops and one *palawija* crop per year. Yields are between 3.5 and 4.5 tons of unhusked rice per hectare, and irrigation water typically is provided by either a semitechnical or a simple irrigation system. Poor water control can lead to drought stress in the dry seasons and flooding in the wet season.

Use of modern inputs, such as HYV seeds and chemical fertilizers, is less common and less intensive than in the higher-productivity systems. In most cases, only urea and TSP are used, and levels rarely exceed 350 kilograms of total fertilizer per hectare. In some off-Java areas, including Agam, long-duration, traditional seed varieties are still planted, especially at higher elevations. These varieties mature in about five months, so farmers cannot grow two rice crops in a year. Farmers in Agam are aware of improved, shorter-duration seeds and in many cases have tried them, but prefer traditional varieties that offer the same or higher yields, better-tasting rice, natural pest resistance, and a moderate price premium over improved varieties.

The field surveys revealed that rice was reasonably profitable in 1987 (Table 4.7) on *sawah* where water is poorly controlled. Because profits were several hundred thousand rupiah lower than in the higher-productivity systems, *palawija* crops were more competitive. Nevertheless, rice remains more profitable on average (Tables 4.7 and 4.8), and nonrice crops typically are grown on poor-control *sawah* only when rice cultivation is not possible or too risky.

Drainage problems prevent cultivation of nonrice crops during the wet season in most of Indonesia's semitechnical and simple irrigation systems. This technical constraint alone eliminates between 50 and 75 percent of the potential crop substitution that could take place on poor-control *sawah*. To be able to plant crops such as corn or soybeans in the wet season, farmers would have to make substantial investments in removing bunds and raising land surfaces above the surrounding land to facilitate drainage (through this process, of course, the land would cease to be *sawah*). Farmers, understandably, are unwilling to undertake such an investment for potential returns lower than or equal to those from rice.

Sample farmers also perceived rice to be less risky than other crops because of the BULOG floor price for rice and the volatility of *palawija*

TABLE 4.7. *Costs and returns for rice, poor-control, irrigated* sawah, *1987*

	Majalengka		Kediri	Pinrang-Sidrap		Agam	
	Wet (Nov–Feb)	Dry (Mar–Jun)	Wet (Nov–Feb)	Wet Nov–Feb)	Dry (Mar–Jun)	Wet lowland improved seeds	Wet upland traditional seeds
Returns (000 rupiahs per hectare)							
Yields (kilograms per hectare)	4,500	4,500	4,250	4,500	3,500	4,000	3,750
Value (rupiahs per kilogram)	175	200	180	160	225	265	290
Gross value	788	900	765	720	788	1,060	1,088
Value of harvest share	158	180	77	72	79	147	150
Value of farmer's share	630	720	688	648	709	913	938
Costs (000 rupiahs per hectare)							
Cash input costs	64	61	71	71	74	52	33
Cash labor costs	256	266	192	50	50	411	391
Cash return	310	393	425	527	585	450	514
Implicit family labor	60	73	21	98	114	52	32
Land rent	125	125	175	298	328	361	415
Implicit interest cost	13	13	11	5	5	19	21
Net return	112	182	218	126	138	18	46

SOURCE: Field surveys.

TABLE 4.8. *Costs and returns for* palawija *crops, dry-season, poor-control, and rainfed* sawah, 1987

	Majalengka	Klaten		Agam	
	Soybeans	Corn	Soybeans	Lowland corn	Upland green beans
Returns (000 rupiahs per hectare)					
Yields (kilograms per hectare)	700	2,500	900	2,500	4,000
Value (rupiahs per hectare)	700	200	700	175	425
Gross value	490	500	630	438	1,700
Costs (000 rupiahs per hectare)					
Cash input costs	68	61	75	65	357
Cash labor costs	187	161	218	334	648
Cash return	235	278	337	39	695
Implicit family labor	23	46	23	122	82
Land rent	125	150	150	200	200
Implicit interest cost	8	7	9	12	20
Net return	79	75	155	−295	393

SOURCE: Field surveys.

yields. Soybeans, in particular, suffer greatly from sporadic pest problems under tropical conditions. Various vegetable crops (e.g., peppers and green beans) provide much higher average profits than rice does, but successive vegetable crops cannot be grown without great risk of severe pest attacks. Given the short duration of most vegetable crops (one to two months), direct competition with rice can be avoided. Finally, survey farmers strongly preferred producing rice for home consumption rather than buying in the market. These additional points, along with higher average profitability, firmly entrench rice as the crop of choice on poor-control *sawah*.

Rainfed Sawah

Rice area on rainfed *sawah* accounts for 20 percent of rice plantings and 18 percent of annual rice production in Indonesia. Rainfed rice is slightly more common in the outer islands but is also prevalent in Java. Like poor-control, irrigated *sawah*, rainfed *sawah* tends to be located in more remote areas that have less supporting infrastructure such as paved roads and agricultural extension.

From an agronomic perspective, the major distinguishing feature of rainfed *sawah* is its dependence on rainfall for water. Since cultivation with HYV seeds is not successful without sufficient rainfall to flood the land for at least two months, rice production on rainfed lands is considerably riskier than that in any of the irrigated systems. Farmers must wait until the heavy monsoonal rains have started or risk drought stress at the crucial early stages of plant growth. But if they wait too long, they risk drought stress late in the crop cycle from an early cessation of the rains.

The lack of water control often reduces yields because of drought stress. Yields rarely exceed 4.5 tons of unhusked rice per hectare on rainfed land and are often much lower, particularly outside Java. Yields are reduced by lower fertilizer levels; applications generally are less than 300 kilograms total per hectare. Pesticides tend to be used when necessary (and often when not) on all wetland rice, including rainfed rice. Improved seed varieties are also planted by most farmers in Java's rainfed areas, though less commonly so off Java.

Since the rainy season lasts on average about five months throughout much of Indonesia, only one rice crop is grown on about 90 percent of rainfed *sawah*. Two rice crops are possible in areas with very long wet seasons (mostly located in Sumatra). Two rice crops also are possible in certain areas if the first rice crop is dry seeded with very short-maturing seed varieties (e.g., IR 36) just before the rains start (*padi gogo rancah*). The spread of dry-seeded *sawah* rice has been limited by low yields, high risk, and inappropriate soils, although the technique has been successfully introduced into several areas of Java. Scope for increasing *padi gogo rancah* area in the future is limited, and such area has recently fallen since the practice was started in the early 1980s.

Table 4.9 indicates that rice cultivation is profitable on rainfed *sawah*. In fact, profits are higher on average (but more variable) than those for the low-productivity, irrigated system. With the exception of dry-seeded rice, crop substitution is a moot issue on rainfed *sawah*. Cultivation of nonrice crops is not feasible in the wet season because of drainage problems. Rice cannot be grown in the dry season because of a lack of water. Survey farmers in Klaten, Central Java, who grow dry-seeded rice, substitute

TABLE 4.9. *Costs and returns for rice, rainfed* sawah, *1987*

	Klaten		Majalengka	Pinrang/Sidrap
	Flooded	*Padi gogo rancah*	Flooded	Flooded
Returns (000 rupiahs per hectare)				
Yields (kilograms per hectare)	4,000	3,500	5,000	3,500
Value (000 rupiahs per hectare)	0.20	0.20	0.175	0.180
Gross value	800	700	875	630
Value of harvest share	53	47	175	63
Value of farmer's share	747	653	700	567
Costs (000 rupiahs per hectare)				
Cash input costs	74	77	59	67
Cash labor costs	223	174	251	50
Cash return	450	402	390	450
Implicit family labor	36	26	36	95
Land rent	150	150	125	260
Implicit interest cost	12	10	12	5
Net return	252	216	217	90

SOURCE: Field surveys.
NOTE: Returns to rainfed rice in Kediri and Agam are very similar to those for low-productivity *sawah* rice.

freely between rice and soybeans before the wet season, according to expected relative profitabilities, but this substitution represents a very small proportion of cultivated area and an even smaller proportion of total production in Klaten.

Conclusion

Indonesia's rice production environment is diverse. This chapter has characterized this diversity in terms of agronomic and economic differences. The aggregate response of wetland rice farmers to possible price and regulatory policy changes is examined in Chapter 7. For government policies that have a country-wide impact, such as the fertilizer subsidies and the floor price for rice, the diversity of rice systems causes farmers to have varying responses, even when they live in the same geographical region. The high degree of commercialization of rice production tends to make responses more homogeneous because large numbers of farmers are responding directly to market forces. But even in fully commercial rice operations, the diversity of soil types, water availability, farmers' skills, and alternative potential crop choices can lead to differing responses to uniform policy changes.

The budget-based analysis provided in this chapter indicates that rice production dominates in Indonesia's wetland areas. Rice offers much higher financial returns than do competing crops in all wetland rice systems. In addition, technical constraints limit the production of nonrice crops in the wet season in many environments. Because limited substitution possibilities are available on already existing rice land, most future growth in cultivated rice area will have to come from investments in irrigation infrastructure, new technologies made available from research, and changes in regulatory policy. This key result, which reflects both technical and economic constraints in rice farming, has important implications for the selection of policies to encourage increased rice production.

APPENDIX 4.1. *Field Budget Assumptions*

Klaten, Central Java

	Good-control *sawah*		Moderate-control *sawah*			Rainfed *sawah*			
	Wet rice	Dry rice	Wet rice	Dry rice	Soybeans	Flooded rice	Dry rice	Soybeans	Corn
Fertilizer (kilograms per hectare)	400	500	400	500	250	350	350	150	500
Liquid insecticide (liters per hectare)	2.5	1.5	3	2	5	4	3	6	–
Granulated insecticide (kilograms per hectare)	16	12	16	12	–	16	16	–	–
Seeds (kilograms per hectare)	35	35	35	35	50	35	50	50	25

GENERAL ASSUMPTIONS

Wages (rupiahs per hour)	
Animal labor	850
Male labor	235
Female labor	130
Planting labor	145
Implicit family	200

Prices	
Fertilizer	110 rupiahs per kilogram
Rice seed	400 rupiahs per kilogram
Soybean seed	800 rupiahs per kilogram
Corn seed	250 rupiahs per kilogram
Liquid insecticide	3,000 rupiahs per liter
Granulated insecticide	650 rupiahs per kilogram

Majalengka, West Java

	Good-control *sawah*			Poor-control *sawah*			Rainfed *sawah*
	Wet rice	Dry rice	Soybeans	Wet rice	Dry rice	Soybeans	Flooded rice
Fertilizer (kilograms per hectare)	400	500	150	400	400	150	350
Liquid insecticide (liters per hectare)	2	1	5	2	1	5	2
Granulated insecticide (kilograms per hectare)	–	–	–	–	–	–	–
Seeds (kilograms per hectare)	35	35	36	35	35	36	35

GENERAL ASSUMPTIONS

Wages (rupiahs per hour)		Prices	
Animal labor	1,000	Fertilizer	110 rupiahs per kilogram
male labor	250	Rice seed	400 rupiahs per kilogram
Female labor	125	Soybean seed	1,000 rupiahs per kilogram
Planting labor	17,500 rupiahs per hectare	Liquid insecticide	3,000 rupiahs per liter
Implicit family	250		
Tractor	70,000 rupiahs per hectare		

Kediri, East Java

	Moderate-control *sawah*						Poor-control *sawah*	Rainfed *sawah*	
	Wet rice	Dry rice 1	Dry rice 2	Hybrid corn	Second-season soybeans	Third-season soybeans	Wet rice	Flooded rice	
Fertilizer (kilograms per hectare)	500	500	500	500	725	150	250	400	400
Liquid insecticide (liters per hectare)	1	1.5	1.5	1.5	–	4	4	1	1
Granulated insecticide (kilograms per hectare)	12	12	12	12	12	–	–	–	–
Seeds (kilograms per hectare)	35	35	35	35	35	50	50	60	60
Irrigation fee (rupiahs per hectare)	–	5,000	7,500	30,000	30,000	–	50,000	–	–

GENERAL ASSUMPTIONS

Wages (rupiahs per hour)		Prices	
Animal labor	650	Fertilizer	120 rupiahs per kilogram
Male labor	225	Rice seed	325 rupiahs per kilogram
Female labor	150	Soybean seed	1,000 rupiahs per kilogram
Spraying labor	250	Arjuna seed	800 rupiahs per kilogram
Implicit family	200	Hybrid seed	1,800 rupiahs per kilogram
		Liquid insecticide	3,000 liters per kilogram
		Granulated insecticide	750 rupiahs per kilogram

(continued)

APPENDIX 4.1 (*Continued*)

Pinrang-Sidrap, South Sulawesi

	Moderate-control *sawah*		Poor-control *sawah*		Rainfed *sawah*
	Wet rice	Dry rice	Wet rice	Dry rice	Flooded rice
Fertilizer (kilograms per hectare)	350	350	250	250	200
Liquid insecticide (liters per hectare)	–	–	2	2	–
Granulated insecticide (kilograms per hectare)	20	20	–	–	10
Seeds (kilograms per hectare)	30	30	30	30	30
Herbicide (liters per hectare)	2	2	1	1	1
Irrigation fee (rupiahs per hectare)	–	3,000	–	3,000	–

GENERAL ASSUMPTIONS

Wages		Prices	
Animal labor	50,000 rupiahs per hectare	Fertilizer	120 rupiahs per kilogram
Tractor	65,000 rupiahs per hectare	Rice seed	300 rupiahs per kilogram
Shadow wage	250 rupiahs per hectare	Liquid insecticide	3,000 rupiahs per liter
Gotong royong	20,000 rupiahs per hectare	Granulated insecticide	750 rupiahs per kilogram
Planting labor	250 rupiahs per hour	Herbicide	6,000 rupiahs per liter

Agam, West Sumatra

	Good-control *sawah*		Moderate-control *sawah*	Poor-control *sawah*		
	Wet/dry rice	Soybeans	Wet/dry rice	Wet rice	Corn	Green beans
Fertilizer (kilograms per hectare	400	250	350	300	400	240
Liquid insecticide (liters per hectare)	2	7	2	1	3	9
Granulated insecticide (kilograms per hectare)	8	–	8	4	–	–
Seeds (kilograms per hectare)	25	40	25	25	40	16

GENERAL ASSUMPTIONS

Wages		Prices	
Animal labor	1,150 rupiahs per hour	Fertilizer	125 rupiahs per kilogram
Planting labor	450 rupiahs per hour	Rice seed	400 rupiahs per kilogram
Implicit family	400 rupiahs per hour	Soybean seed	800 rupiahs per kilogram
Harvest cutting	400 rupiahs per hour	Corn seed	175 rupiahs per kilogram
		Bean seed	2,500 rupiahs per kilogram
		Liquid insecticide	3,000 rupiahs per liter
		Granulated insecticide	700 rupiahs per kilogram
		Tractor rental	80,000 rupiahs per hectare
		Sprayer rental	1,000 rupiahs per day

5. The Rural Labor Market in Indonesia

Rosamund Naylor

An understanding of the rural labor market is central to an analysis of the effects of past and prospective rice policies in Indonesia. Labor remains the most important input in rice production, and rice traditionally has been the largest source of rural employment on Java. But the relative importance of rice has diminished rapidly in the past two decades, and the effects of rice policy on the rural labor market are much different today than in earlier periods. Just how different they are depends on the efficiency with which the rural labor market functions.

The extent of integration between rice and other rural labor markets, and between rural and urban labor markets, influences the changes in employment and wage levels that would result from changes in rice policy. If the labor market for rice is isolated from the rest of the economy, changes in profitability of rice production can be expected to have a direct impact on employment and wages in the rice sector. The large size of the rice sector in the Indonesian economy ensures that such changes would have major significance for aggregate employment and rural incomes. If, on the other hand, the labor market is well integrated, changes in rice employment can be accommodated with relative ease by other parts of the economy; farmers would have difficulty reducing real wages paid in rice production for fear of losing workers to other pursuits.

The efficiency of rural labor markets in Indonesia is an issue of considerable controversy. Studies undertaken at different times create such contrasting scenarios that it is difficult for the reader to believe that the various authors are considering the same country. Analyses also have been plagued by a lack of reliable data so that no consensus has emerged about such fundamental phenomena as the direction and magnitude of change in the level of rural wages. This chapter reassesses the structure of rural labor markets with attention focused on two principal issues—competitiveness in the market for unskilled labor and real wage trends.

The first section of the chapter presents evidence from field surveys conducted in 1987–88 on off-farm job opportunities and relative wages (off-farm versus on-farm) for unskilled rural workers. The goal of this work is to investigate whether off-farm jobs are widespread, accessible, and well paid relative to those in rice production. Attention also is given to the selection of employment during the course of the year to understand the role of the rice economy in providing full employment for unskilled laborers. The next section examines recent changes in labor-hiring institutions and work practices in rice production to determine how cultural arrangements evolve in relation to demand and supply conditions in the rural and urban labor markets. The third section provides an analysis of real wage trends in rice production on Java between 1976 and 1988. These trends are evaluated in the context of the observed changes in employment opportunities.

Alternative Views of the Rural Labor Market

Before the introduction of Green Revolution technologies in the late 1960s, the rice sector on Java was characterized as a subsistence sector, capable of absorbing an expanding labor force but not of stimulating rural economic growth (Geertz 1963). Significant gains in rice production and land productivity resulted from increasingly intensive cultivation practices, but at best these changes maintained per capita income levels: "The ever driven wet-rice village found the means by which to divide its growing economic pie into a greater number of traditionally fixed pieces and so to hold an enormous population on the land at a comparatively very homogeneous, if grim, level of living" (Geertz 1963, p. 100).

Geertz's equally dismal outlook for the future was premised on a continuous growth of the rural population (and the rural labor force) within Java's confined productive land area and on little expansion of off-farm employment. Until 1970, few productive employment opportunities were available to unskilled workers in the nonagricultural sectors of the rural economy.

Labor market surveys conducted in the 1970s, reported by White (1976, 1979) and Hart (Hart and Sisler 1978; Hart 1986a, 1986b), found that nonagricultural job opportunities for unskilled workers in rural areas were limited in scope and earning potential and that employment in rice production did not rise appreciably with higher rice output. These conditions led to monopsonistic behavior on the part of landowners and increasing income polarization between landholders and landless laborers. Field surveys from the 1970s also supported the existence of a segmented labor market, evidenced by widely different wage rates within the same village for a given age-sex group and season (Lluch and Mazumdar 1985). Hart presented evidence that this divergence in wage rates was attributable to preferential hiring arrangements within the rice sector based on income and social status (Hart and Sisler 1978; Hart 1986b).

Empirical investigations in the 1980s depict a markedly different pattern. Workers earn comparable wages in agricultural and nonagricultural activities. In many areas, labor moves easily between sectors and regions of the economy in response to relative wage differences and the seasonal availability of work opportunities. Research conducted in the mid-1980s by Collier et al. (1988) and Manning (1986a, 1986b, 1988) concluded that an expanding range of employment opportunities, brought about by high rates of economic growth in both rural and urban locations, improved the bargaining position of unskilled workers and helped to distribute income gains more equitably throughout the economy. These analysts found the labor market for unskilled workers on Java increasingly well integrated. The extent to which this transition in rural labor markets has occurred on and off Java remains a source of debate among researchers in Indonesia.

Recent Empirical Evidence on the Rural Labor Market

Field survey evidence presented in this section is used to describe the income-earning activities available to unskilled workers.[1] This evidence helps to establish whether the process of marginalization—the movement of unskilled labor from the agricultural sector into low-paying and low-productivity jobs in the nonagricultural sector—is prevalent in rural Java or whether, instead, unskilled laborers are able to work in nonfarm jobs at wages comparable to those earned in agriculture. These results, in turn, show the extent to which landowners are able to exert monopsony power over landless laborers. In other Asian countries, the emergence of a wide range of equally remunerative employment opportunities for unskilled laborers in nonagricultural activities has provided convincing evidence of

[1]The rural labor market survey was part of a larger survey of rice production technology, crop substitution, and rural labor markets associated with rice cultivation conducted by Heytens and Naylor in 1987–88. The survey of rural labor markets consisted of interviews with rice farmers, farm laborers, owners of rural industries, and unskilled workers in rural industries. The numbers of respondents in the sample were as follows:

Kabupaten	Villages	Farmers	Unskilled laborers[a]	Industries	Other
Kediri	15	28	25	9	3 *penebas*[b]
Klaten	16	31	39	19	2 *penebas*
Majalengka	20	41	42	7	2 village leaders
Pinrang	12	26	8	5	many local
Sidrap	11	22	10	4	villagers

[a]On and off the farm.
[b]See note 2.

Because the survey sites were chosen to emphasize dominant areas of rice production, they are biased slightly toward areas of strong economic growth. Few remote upland regions with poor access to large cities or towns were included in the survey.

a transition from monopsonistic to competitive labor market conditions (Bardhan 1984; Squire 1981).

The survey results confirm previous findings that most rural laborers on Java pursue multiple income-earning activities, within a given day, within a season, and over the course of a year (Collier et al. 1982b, 1988; White 1986; Manning 1986a, 1986b). Regions vary in employment opportunities, and the survey results identify several distinct patterns in labor market behavior. Both institutions and workers' behavior are remarkably flexible in accommodating the pursuit of full employment. But in general, labor mobility is greatest between the urban informal and rural labor markets and between the agricultural labor market and the off-farm rural labor market (Rucker 1985). Construction activity in the cities and villages has been one of the most significant forces attracting male laborers from the agricultural sector since the early 1980s (Manning 1986b, 1988).

Job opportunities for landless laborers in rice production in any given season depend on the proportion of *sawah* planted in paddy. In the rainy season (November to February), when virtually all of the *sawah* is in paddy, laborers work in agriculture for two months on average. The number of labor days employed in rice production is higher for workers who harvest with contract teams (*tebasan*[2]) and for those in areas where irrigation schedules stagger the planting and harvesting cycles for farms throughout the season. In the dry season, employment in rice production varies considerably by region. The diversity of production environments in the survey sites thus provides a broad basis for evaluating labor market mobility between sectors and regions on Java.

Three crops of rice are grown on one-third of the *sawah* in Klaten (Central Java), and laborers work for one to two months on average during each of the dry-season plantings (March to June and July to October). In Majalengka (West Java), paddy is planted twice a year in the irrigated *sawah* and once a year in the rainfed *sawah*, but rice is rarely triple cropped because of insufficient water. Laborers are able to find employment in the rice sector for roughly two months in each of the rice seasons. Many look for work outside of the agricultural sector in the second dry season. Rice is triple cropped on only 5 to 10 percent of the *sawah* in Kediri (East Java). Total days worked in agriculture per laborer fall from fifty to one hundred days in the rainy season to twenty-five to forty days in the second dry season, when only a small portion of the *sawah* is in paddy.

The substitution of *palawija* crops for paddy in the dry season reduces the demand for labor in the agricultural sector. Labor inputs for soybeans

[2]*Tebasan* is a contract arrangement by which a farmer sells a standing crop of paddy to a labor contractor (*penebas*), who hires a team of workers to harvest the *sawah*.

and corn are only 50 to 75 percent of those for rice in the dry season (Table 5.1). In Kediri, where corn is grown extensively in the *sawah* during both dry-season plantings, laborers are able to work in corn production for fifteen to thirty days per planting. Soybeans and vegetables are the predominant *palawija* crops in the second dry season in Majalengka. Laborers remaining in the village work for twenty to forty days per season in the production of these crops. Sugar and tobacco are labor-intensive crops, but less so than paddy. In Klaten, government sugar and tobacco programs permit male laborers to work for three to five months per year in sugar production and male and female laborers to work for up to seven months per year in the tobacco fields.

Interviews with farm laborers and workers involved in various off-farm occupations indicate that they engage in multiple-earning activities during the course of a year or within a season rather than within a given day. Casual and self-employed activities such as informal trading, driving *becaks* (rickshaws), and making or carrying bricks are the main off-farm activities pursued within a given season. Employment of longer duration, mostly work on infrastructure or construction projects, is more prevalent on a seasonal basis (i.e., during the dry season).

Employment Patterns in Kabupaten Kediri (East Java)

Interviews with unskilled workers in a wide range of rural activities show that labor supply is allocated on the basis of relative wages (including wage equivalents of meals), job duration, physical exertion, perceived risk, and family life-style. Table 5.2 reports a sample hierarchy of employment opportunities for unskilled labor based on survey interviews in Kediri in 1987. At the top of the hierarchy are jobs outside of the agricultural sector which offer significantly higher pay and benefits on a semipermanent basis. For example, road work on a piecework basis pays an average of Rp 3,000 to 4,000 per day for a five- to six-month period, compared to the average wage for males in agriculture of Rp 1,900 per day (meals included) for the same number of hours (eight hours per day). These off-farm activities generally require strenuous physical exertion or some training on the job. Because barriers to entry into this class of activities are rare for unskilled workers, these jobs contribute to higher wage levels in rural areas.

Below these high-paying activities are jobs that pay approximately the same wage as agriculture on a daily basis but offer long duration of several months or even years. Within this group, employment preference between agriculture and other jobs is mixed. For example, a construction worker, who could earn Rp 1,500 per day on a six-month basis, preferred agricultural work because he received meals in addition to the same daily wage. Third in the hierarchy are jobs that offer a higher daily wage than

TABLE 5.1. *Labor inputs for paddy and* palawija *(hours per hectare)*

KABUPATEN KLATEN

	High-productivity *sawah*			Low-productivity *sawah*			
	Wet	Dry	Tobacco	Wet	Dry	Soybeans	Corn
Land preparation	260	260	1,176	260	260	390	300
Planting	250	250	144	250	250	300	180
Weeding/fertilizing	472	460	546	531	516	348	204
Water management	100	175	228	125	200	–	–
Harvesting	400	400	126	350	350	150	300
Dry	24	16	324	20	12	88	228
TOTAL	1,506	1,561	2,544	1,536	1,588	1,276	1,212

NOTE: Tobacco is a two-season crop, May–December.

KABUPATEN KEDIRI

	Paddy			Corn	
	Wet season	First dry season	Second dry season	Traditional	Hybrid
Land preparation	140	140	140	140	140
Planting	250	250	250	60	60
Weeding/fertilizing	310	375	375	190	190
Water management	100	425	425	50	50
Harvesting	450	450	500	175	200
Dry	30	12	12	60	70
TOTAL	1,280	1,652	1,702	675	710

NOTE: Dry includes cleaning and processing for soybeans.

KABUPATEN MAJALENGKA

	High-productivity irrigated *sawah*, paddy (tractor)		Low-productivity irrigated *sawah*, paddy (animal)		Rainfed *sawah*	Irrigated *sawah*, soybeans, dry
	Wet	Dry	Wet	Dry		
Land preparation	162	162	455	455	600	259
Planting	250	250	250	250	250	130
Weeding/fertilizing	369	404	469	504	469	249
Water management	100	175	125	200	30	
Harvesting	800	800	750	750	750	140
Dry	24	16	20	12	20	140
TOTAL	1,705	1,807	2,069	2,171	2,119	918

KABUPATEN PINRANG-SIDRAP

	Medium-productivity irrigated *sawah*, paddy (tractor)		Low-productivity irrigated *sawah*, paddy (animal)		Rainfed *sawah*
	Wet	Dry	Wet	Dry	
Land preparation	18	18	45	45	60
Planting	100	100	100	100	100
Weeding/fertilizing	36	36	71	71	65
Pest/insect	70	120	70	150	60
Water management	300	280	280	270	270
Harvesting	24	8	24	8	16
TOTAL	548	562	590	644	571

SOURCE: Field surveys.

TABLE 5.2. *Employment hierarchy, Kabupaten Kediri, 1987*

Activity/industry	Wage[a]	Sex	Hours per day	Comments
Paddy production	Rp 1,200	M	5	Seasonal
	Rp 700	F	4	Seasonal

Class 1: Permanent or semipermanent employment. Daily wage exceeds the wage in the rice sector.

Activity/industry	Wage[a]	Sex	Hours per day	Comments
Sugar mill	Rp 60–80,000 per month (minimum)	M	8	Plus benefits; seasonal: May–December
Construction/ infrastructure	Rp 1,500–2,000 per day (minimum); Rp 3,000–4,000 per day (average)	M	8	Piecework; project basis, 5–6-month duration

Class 2: Permanent or semipermanent employment. Daily wage is similar to the wage in the rice sector.

Activity/industry	Wage per day[a]	Sex	Hours per day	Comments
Vulcanizing plant	Rp 1,200–1,500	M	8	Plus benefits; year-round
Corn seed company	Rp 1,200–1,450	M	8	6 months per year: June–December
Large cigarette factory[b]				
Construction	Rp 1,500	M	8	Project basis, 5–6 months per year
Rolling cigarettes	Rp 880–1,770	F	8	Plus benefits; piecework; year-round

Class 3: Seasonal or part-time employment.
Daily wage often exceeds the wage in the rice sector.

Activity/industry	Wage per day	Sex	Hours per day	Comments
Selling *krupuk*[c] in market	Rp 5,000–7,000	M	11	
Carrying heavy loads (*tukang keras*)	Rp 2,000	M	8–10	Strenuous
Tailor	Rp 3,000–6,000	M, F	5	Part time

Class 4: Seasonal or part-time employment. Wage is on par with the wage in the rice sector. Constant movement between these jobs and work in the rice sector.

Activity/industry	Wage per day[a]	Sex	Hours per day	Comments
Becak driver	Rp 800–2,500	M	8–10	Net of rent
Fruit/vegetable trader	Rp 600–2,000	M, F	6–8	
Street food vendor	Rp 800–1,500	M	8	Net of costs
Maid in village	Rp 1,500	F	10	Includes three meals per day

Class 5: Permanent employment.
Wage is much lower than the wage in the rice sector.

Activity/industry	Wage per day[a]	Sex	Hours per day	Comments
Krupuk factory[c]	Rp 750	F	10	Fulltime; no meals; mostly young females
Small cigarette factory (low quality)	Rp 500–1,000	F	8	Full time; piecework

SOURCE: Field surveys.
[a]Includes meals.
[b]Gudang Garam cigarette factory.
[c]Food product made of cassava starch and additives, often shrimp.

agriculture but are not guaranteed on a daily basis. Although respondents in these activities clearly preferred nonagricultural work because of the greater physical exertion required for farm labor, they continued to work on their own land and for other farmers in all seasons on a part-time basis.

The fourth class of job opportunities is the most representative for landless laborers on Java throughout the year. This class of self-employed workers includes *becak* drivers and informal traders. Earnings are comparable to those for nonfarm activities and agricultural work, and labor moves smoothly between these sectors within each season. These off-farm activities are not marginal in the sense of having inferior quality or low income-earning potential, and they are often preferred to agricultural work. On an hourly basis, wages are roughly the same as those in agriculture. Within this category, the opportunity cost of agricultural labor generally equals the wage rate. The constant flow of workers between the farm and nonfarm sectors for this category of jobs is evidence of a high degree of integration within the regional labor market.

The final category of employment opportunities consists of jobs that are clearly inferior to agricultural labor in both wage and quality of work but offer a means to employ family members with few work alternatives and low opportunity costs to the family. This category is made up of formal sector jobs that can absorb labor on a full-time or seasonal basis. Although these jobs offer permanent wages, working conditions are poor, hours are long, and meals are not included. Most of the laborers in this group are young women who often return to their villages to take part in rice harvests.

Employment Patterns in Kabupaten Majalengka (West Java)

The pattern of off-farm employment opportunities varies between regions. Fewer off-farm employment opportunities exist for unskilled labor in Majalengka than in Kediri. As in much of West Java, widespread migration to the major cities of Jakarta and Bandung occurs on a seasonal basis (Manning 1986b). Information and transportation between rural and urban areas is good, and housing generally is provided by employers, often construction firms, in the cities. Earnings are higher in the cities than in the villages. The migrants are mostly male, causing the role of women in the rural labor force to be more pronounced than in East Java (Sjahrir 1990).

Village workers tend to migrate to the cities in groups. In some cases, 70 percent of the unskilled laborers routinely leave the village together to work on a single construction project. In other villages, virtually no seasonal migration to the urban areas occurs. Information about work opportunities in urban areas generally flows by word-of-mouth, although occasionally a person from a project in the urban area is sent to the rural areas to recruit workers. Migrants leave the village either for an extended period (three to five months) or for several shorter periods (one to two months) during the year.

Although emigration is a prominent part of rural employment patterns in Majalengka, opportunities also exist locally (Table 5.3). Tile factories are the principal nonagricultural activity in the region. The larger tile factories operate throughout the year and hire two hundred daily workers (men, women, and children) on average. To secure an adequate supply of daily laborers, the large factories send trucks to the villages to pick up those who wish to work that day and deliver them back to the villages after work. The small factories hire forty daily workers on average, who come on their own from nearby villages. Entry into the industry is unrestricted. Wages in the tile factories are similar to those in agriculture. Employment in local construction also is prevalent during the dry season, as are a variety of informal jobs such as making bricks, digging and carrying sand from the rivers for construction projects, and collecting firewood.

Employment Patterns in Kabupaten Klaten (Central Java)

Klaten is the most densely populated of the survey sites. Supplies of labor are ample, and the likelihood of an "involuted" labor market rife with market failures appears greater than in the other sites. Instead, a very numerous and diverse set of demands for unskilled labor has evolved within the Klaten labor market. The nonfarm sector in this region is dominated by a wide range of small-scale industries such as food processing, textiles, batik, furniture and plywood manufacturing, and ironworks. Table 5.4 shows the number of industries and employees by size classification which were in operation in 1986. Most of the industries are household industries, and demand for unskilled labor by this group is much higher than demand for labor from the larger industrial groups. Most household industries are not mechanized; inputs consist primarily of labor and raw materials.

Several farm laborers interviewed in Klaten work in construction for one to four months during the dry season either within their villages or in the nearby cities of Solo or Yogyakarta. Many others work in small rural industries on a part-time basis when they are unable to find work in the *sawah*. Interviews with unskilled laborers and owners of industries confirm a high degree of labor mobility between the agricultural and nonagricultural sectors of the rural economy. Many of the unskilled employees of the industries return to the *sawah* to participate in the paddy harvests. Others work in a variety of tasks in the cultivation of paddy as well as in industry on a part-time basis. All of the laborers interviewed had

TABLE 5.3. *Employment hierarchy, Kabupaten Majalengka, 1988*

Class 1: Full-time employment

Activity/industry	Wage per day[a]	Sex	Hours per day	Comments
Large rice mill	Rp 8,000–10,000	M	15	Busy season
	Rp 3,000–5,000	M	8	Slow season; Year-round: one day on, one day off
Small rice mill	Rp 2,000	M	8	Year-round
Sugar mill	Rp 75,000–90,000 per month	M	8	May–October
Sweet potato factory	Rp 2,000	M	8	Year-round

Class 2: Seasonal employment

Activity/industry	Wage per day[a]	Sex	Hours per day	Comments
Paddy production	Rp 2,000–2,500	M	7	Seasonal
	Rp 1,000	F	6	
Drive tractor	Rp 3,000	M	7	
Local construciton	Rp 2,500	M	8	4–6 months per year
Local infrastructure	Rp 3,500	M	8	Project basis
Construction in cities	Rp 2,500–4,500	M	8	No meals
Laborer in cities	Rp 2,500–5,000	M	8	No meals

Class 3: Temporary seasonal jobs (supplemental income)

Activity/industry	Wage per day[a]	Sex	Hours per day	Comments
Tile factory	Rp 1,000–1,500	M, F	8	No meals
Making bricks	Rp 1,500–2,000	M	8	No meals
Collecting sand	Rp 1,500–2,000	M	8	No meals
Collecting firewood	Rp 1,000–1,500	M	6–8	No meals
Laborer in sugar mill	Rp 1,500–2,000	M	8	No meals
Becak driver	Rp 900–2,500	M	10–12	Net of costs
Laborer in sub-DOLOG warehouse	Rp 2,500	M	6	No meals

Source: Field surveys.
[a]Includes meals.

TABLE 5.4. *Industries in Kabupaten Klaten, 1986*

Type of industry	Household	Small	Medium	Large
Food, drink, tobacco	6,922	395	221	21
Textile and clothing	1,984	582	54	1
Raw materials and construction products	4,580	239	2	–
Handicrafts	5,561	31	16	–
Ironworks	1,766	267	–	–
Other	314	49	3	–
TOTAL	21,127	1,563	296	22
TOTAL EMPLOYEES	43,225	14,336	9,404	3,191

SOURCE: Kantor Statistik Kabupaten Klaten (1986).

knowledge of a range of employment opportunities and usually allocated their time to the most remunerative activity available. Table 5.5 lists the industries surveyed in Klaten and the wages paid for each activity.

Nonagricultural wages for unskilled laborers in factories are often lower than agricultural wages. Factory work is attractive because employment is full time or continuous for many months, the work is less physically

TABLE 5.5. *Industries surveyed in Kabupaten Klaten, February 1988*

Industry	Wage per day[a]	Sex	Hours per day	Comments
Infrastructure Projects	Rp 1,250	M, F	6.5	Daily wage
	Rp 2,000–3,000	M, F	6.5	Piecework
Tobacco factory	Rp 650–800	F	7	Seasonal
Petty trade	Rp 500–1,000	M, F	10	
Construction	Rp 1,500–3,000	M	8	
Tailor	Rp 2,500	M, F	10	Part time
Furniture	Rp 2,500–3,500	M	8	Unskilled
	Rp 6,500	M	8	Skilled
Textile, government firm	Rp 1,500	M, F	8	Plus benefits
Textile, private firm	Rp 500–1,500	M, F	7	Piecework
	Rp 1,000–1,500	M, F	7	Daily wage
Ironworks	Rp 1,300	M	7	
Batik, Solo	Rp 2,000–5,000	M, F	8	
Becak, Solo	Rp 2,500–5,000	M	10	
Pharmaceutical	Rp 825–1,200	M, F	7	Plus benefits
Bread, snacks	Rp 800–1,500	M, F	8	
Plywood	Rp 1,250–1,550	M	8	Plus benefits
Cotton wool	Rp 650–1,000	M, F	8	Daily wage
	Rp 750–1,500	M, F	8	Piecework
Calcium carbonate	Rp 1,600–2,100	M	8	Plus benefits
Woor (ingredients for	Rp 250–800	F	5	
clove cigarettes)	Rp 1,000–1,500	M	8	
Publishing firm	Rp 1,000–2,000	M, F	7	
Becak driver, Klaten	Rp 1,000–2,500	M	10	

SOURCE: Field surveys.
[a]Includes meals.

difficult, and the value of the product is not high (e.g., bread factory, tobacco factory, and *woor* [ingredients for clove cigarettes] manufacturing). Wages are the same or higher than agricultural wages for piecework labor and for physically difficult work, especially work on construction projects. For seasonal work in infrastructure projects, labor is taken directly from the agricultural sector, and contractors base their wage rates on local agricultural wages. Wages for unskilled laborers in the larger factories, such as the government textile firm and the publishing firm, are similar to those in agriculture, but benefits are provided and the work is often full time. For *becak* driving, the hours worked per day are long, but the intensity of work effort is less than that demanded of agricultural labor. For the most part, therefore, wage differentials are easily explained by factors other than market failures—duration of employment, difficulty of work, and the value of output.

The Rural Labor Market in Kabupatens Pinrang and Sidrap (South Sulawesi)

Employment activities differ greatly between rural areas on and off Java. South Sulawesi was chosen as an off-Java survey site to provide a contrast with the rural labor market on Java. The results of the survey in South Sulawesi cannot be generalized for all areas outside of Java, but they do indicate broad employment patterns, characteristic of parts of Indonesia that have a much higher ratio of land to labor than does Java.

More than half of the *sawah* in Pinrang and Sidrap is now double cropped in rice. Large public sector investments in technical irrigation systems during the 1970s transformed a majority of the *sawah* in the region from rainfed to irrigated rice land. Agroclimatic conditions and labor constraints prevent farmers from planting paddy three times a year. *Palawija* crops are rarely grown in the *sawah;* low yields and the lack of adequate processing and distribution facilities in the area cause these crops to be much less profitable than rice. Rice varieties have a long growing cycle (IR 42 seed variety with 135 days to maturity), and there is usually a fallow period of a month or more between crops.

Labor inputs in paddy production in Pinrang-Sidrap are significantly lower than on Java. This difference arises because of the relative scarcity (and higher cost) of hired labor in South Sulawesi. Most families in Pinrang and Sidrap have access to land through purchase or sharecropping arrangements. The number of landless laborers available for hire is relatively low, and women of Buginese (an ethnic group native to South Sulawesi) origin do not participate in rice production apart from harvesting. The shortage of labor available to the agricultural sector in the region has led to the widespread use of family labor and *gotong-royong* (mutual unpaid labor exchange between friends and neighbors) for most tasks in rice cultivation.

The use of hired labor for planting and harvesting has increased since the mid-1980s. Most of the hired laborers are Javanese migrants who live in Polmas, a *kabupaten* neighboring Pinrang and Sidrap. Javanese women, as well as men, hire themselves out for planting, whereas preharvest workers of South Sulawesi origin, both hired and *gotong-royong*, are all male. The availability of migrant workers originating from Java has been essential to break the labor constraint during peak periods of demand in many areas. The proportion of migrant workers employed in rice production varies enormously by village depending on location and accessibility.

Virtually all of the male farmers interviewed in Pinrang-Sidrap work exclusively on the farm—in the *sawah* for most of the year when paddy is grown and in gardens during the fallow seasons. Although most women do not work outside of the household except during paddy harvests, a variety of home industries—primarily sewing and producing *sutras* (traditional dress) for local and regional markets—contributes to household income. Evidence of multiple earning activities on and off the farm, characteristic of the labor market on Java, is clearly lacking.

Many of the off-farm activities employing hired labor in Pinrang-Sidrap are closely connected to the rice sector. Examples include irrigation expansion and improvement projects, the manufacturing of bags for harvested and milled rice, retail stores for tractors, seeds, and other inputs, and private rice mills. Many of these activities hire local (mostly male) labor, when paddy is not planted, for wages equal to or higher than those earned in rice production. Table 5.6 shows a sample of employment opportunities and wages outside of rice production.

The majority of off-farm employment opportunities are related to the rice sector either directly through production linkages or indirectly through expenditures of rice earnings. For example, home building and home improvements, activities that employ many men from both the villages and the larger towns in the fallow season, are largely dependent on farm incomes generated from the previous rice harvests. Home industry employment for women, such as sewing, is also tied to local expenditures after the harvests.

Employment opportunities unrelated to the rice sector are less significant. A small percentage of the population in each village goes on *merantau* (work outside of the region for long periods of time, e.g., in Kalimantan or Malaysia). The proportion of the population on *merantau* has diminished since technical irrigation became widespread in the region and increased employment and income opportunities in rice production became available.

TABLE 5.6. *Off-farm employment, Kabupatens Pinrang and Sidrap, South Sulawesi, 1987*

Activity/industry	Wage per day[a]	Sex	Hours per day	Comments
Class 1: Activities related to rice production				
Private rice trader	Rp 5,000	M	8	
Laborer in rice mill	Rp 4,000–5,000	M	8	Large mill
	Rp 2,000–3,000	M	8	Small mill
Field laborer, carry sacks from field	Rp 3,000–4,000	M	8	Seasonal
Tractor driver	Rp 10,000	M	5	Seasonal
Rice bag factory	Rp 2,200–2,500	M, F	7	Year-round
Irrigation project	Rp 2,500–3,000	M	8	Project basis, 5–6 months per year
Paddy production	Rp 1,000–1,500	F	7	Locals and migrants
	Rp 2,000–2,500	M	7	
Class 2: Activities related to local expenditures				
Mason, skilled	Rp 5,000	M	8	
laborer	Rp 2,000–3,000	M	8	
Carpenter, skilled	Rp 5,000	M	8	
unskilled	Rp 2,000	M	8	
Brick maker	Rp 2,000	M	8	
Brick seller	Rp 4,000	M	8	
Tailor	Rp 2,000–4,000	M, F	6	
Traditional dress-maker	Rp 1,000–4,000	F	6	
Selling fruit/vegetables in market	Rp 1,500–2,000	M, F	8	
Fishing	Rp 5,000–10,000	M	8	
Becak driver	Rp 1,500–3,000	M	8	
Emigrant remittances	Rp 2,000–3,000	M	8	Extended periods away from home

SOURCE: Field surveys.
[a]Includes meals.

Comparison of Rural Labor Market Conditions on Java and in South Sulawesi

Recent field survey evidence on employment and wages in the rural labor market on Java does not support the hypothesis that there is a labor surplus economy in which job opportunities are limited and wages remain at a minimum subsistence level (Geertz 1963). Evidence from the 1980s depicts a largely competitive rural labor market in which unskilled laborers move between the agricultural and nonagricultural sectors of the economy on a seasonal basis to find employment at comparable wages.

The wide range of employment opportunities available to unskilled workers provides evidence that the ability of landowners to exert monopsonistic power over landless laborers is much weaker than that observed by Hart in the 1970s (1986a, 1986b) and has disappeared in many areas.

Agricultural labor is neither at the top or the bottom of the hierarchy of potential employment opportunities available to unskilled laborers on Java. The notion of landless laborers being pushed out of agricultural jobs into marginal activities with lower hourly wages is not supported by recent empirical evidence. A more plausible explanation is that unskilled labor is being pulled into the rapidly growing nonagricultural sector. Some agricultural workers have left the sector for full-time work, but many have increased part-time employment while maintaining their presence in agriculture. This growth has greatly diminished underemployment during seasons of slack agricultural demand. A large proportion of the laborers interviewed prefer to remain in agricultural work when they have adequate opportunities and can earn returns comparable to those in off-farm work.

The opportunity cost for unskilled laborers in rice production appears to be about equal to the wage rate paid in the agricultural sector. During the slack agricultural seasons, earnings from construction, *becak* driving, selling fruits and vegetables in the market, or employment in a variety of other formal and informal activities are on par with the agricultural wage rate. Labor demand in the rice sector remains important for rural income distribution and employment on a seasonal basis. But broader-based growth in rural incomes, which creates the demand for labor-intensive rural construction and services, and the expansion of employment opportunities in urban areas have become essential elements of the demand for unskilled labor in the rural economy.

Generalizations about labor market behavior on Java are not applicable to many regions off Java. In the South Sulawesi survey site, for example, the rural labor force is more dependent on the rice sector as a source of income and employment than are the areas surveyed on Java. Fewer off-farm employment opportunities exist in rural areas in South Sulawesi, and greater access to cultivable land for most villagers in Pinrang and Sidrap mitigates the need to pursue multiple-earning activities within each season. Although unskilled workers do migrate between islands in Indonesia, the rural labor markets between regions on and off Java are not yet completely integrated.

Institutional Changes in Labor-Hiring Practices

Adjustments in traditional employment practices in rice production offer further evidence of increased competitiveness within the rural labor market. The changing role of cultural practices in determining employment and wage levels for unskilled workers within the rice sector is examined in this section. The premise is that labor-hiring institutions evolve in response to long-term shifts in labor demand and supply. In particular, cultural practices that serve to divide work opportunities among the largest possible number of laborers give way to piecework, cash payments, and contract hiring of labor teams when labor becomes scarce.

Labor-Hiring Institutions

Wide regional diversity in labor-hiring and payment arrangements exists within the Indonesian rice economy, both on and off Java. Existing labor-hiring arrangements include *gotong-royong*, *bawon*, *ceblokan* and *kedokan*, and *tebasan*. Although the specific terms of these arrangements may differ among regions, or even within villages, their structure is largely consistent throughout the rice economy.

Gotong-royong is the mutual exchange of unpaid labor between friends and neighbors within and outside of agricultural production. A reciprocity agreement exists in the institution of *gotong-royong*, but apart from meals no wages are paid for work performed. Within the rice sector, friends and neighbors work together on cultivation tasks and rotate between their farms until the job is completed for each farm. *Gotong-royong* is used most commonly for labor-intensive preharvest tasks such as planting.

Under the *bawon* system, anyone who wishes to harvest may do so (at least within a given region or village). Workers come to the farmers' fields on the day of the harvest, help with the harvesting, and divide among themselves a standard share of their harvest as a payment in kind. This share, usually between one-tenth and one-fifth of the amount harvested, differs between regions according to local scarcity of labor. By adapting to local labor market conditions, the *bawon* system ensures that returns per worker are lowest in regions of most ample labor supply. Incomes are thus spread among the largest possible number of workers in each region.

Ceblokan refers to the hiring arrangement for planting paddy in which a female laborer receives no payment for the task of planting, but she has exclusive access to the harvest and often earns a greater share of the harvest than under the traditional open-harvest system. The *kedokan* system is similar to the *ceblokan* system but includes additional tasks such as weeding. Although husbands and other family members are permitted to help the women with the harvest, the full wage share of the harvest is paid to the woman who performed the preharvest tasks. To ensure job security, female laborers are willing to defer receipt of wage income until after the harvest.

The *tebasan* system is a method of harvesting in which the farmer sells for cash the standing crop of paddy to a labor contractor (*penebas*) before the harvest. The *penebas* hires his own workers, often a group of laborers from his village who travel throughout the region harvesting for the same contractor.

Changes in Labor-Hiring Institutions

The prominence of traditional labor-hiring practices is influenced by local labor demand and supply conditions. But evidence from field surveys reveals considerable changes in hiring practices for both harvest and preharvest tasks in rice cultivation. For example, hiring patterns in Majalengka, where the system of *ceblokan* for planting has been used for many years, are beginning to change in response to regional labor shortages. In most villages surveyed, the farmers still use the *ceblokan* system; if there is abundant labor, the women are willing to plant and wait for the returns from the harvests. In areas of scarce labor, however, wages are paid in addition to the one-sixth to one-fifth share of the harvests for *ceblokan* workers. In the village of Kertaninagin (Kecamatan Kertajati), farmers began paying wages in 1985; 80 percent of the farmers paid wages in the first year, and now all farmers pay wages.

The *ceblokan* and *kedokan* systems of hiring labor for preharvest tasks still are used in some areas of Kediri, although cash wages have largely replaced these institutions for planting and weeding. Evidence of the decline in the use of the *ceblokan* and *kedokan* systems in Majalengka and Kediri is consistent with the empirical results of micro-studies in other areas of Java (Mintoro et al. 1984). In areas of seasonal labor shortages, caused by men migrating to cities or men and women working in rural off-farm activities, the institutions of *ceblokan* or *kedokan* have been replaced by wages. In villages with less access to off-farm work, *ceblokan* and *kedokan* continue to be used.

The choice of harvesting method and the share paid to harvesting workers similarly reflect local conditions in the labor market. In many parts of Java, the use of contract teams of labor for harvesting under the *tebasan* system has replaced the traditional *bawon* system of harvesting. In most cases, the *tebasan* system for harvesting is used to break a seasonal cash constraint of farmers (e.g., for education expenses) or to facilitate the organization and management of harvesting labor. Because of the growth in employment opportunities, the advantages of the *bawon* system in distributing income among the largest numbers of workers have become less important. The harvest share received by the workers on a *tebasan* team generally is lower than that paid to labor under the *bawon* system, but the work is more likely to be full time and thus provides a larger total income to the worker.

The use of contract labor arrangements with wages paid on a piecework basis also has become prevalent for certain preharvest tasks. In most of the villages surveyed in Klaten and Kediri, labor for planting is hired on a contract basis to ensure the farmer of adequate labor supplies at peak periods. The contract cost for hiring planting labor is generally higher than the cost of paying daily wages to individual workers. The piecework wage received by planting workers under the contract system allows farmers to pay laborers according to their productivity.

The evolution of labor-hiring practices in rice production in Pinrang and Sidrap has differed from that in the survey areas on Java. Many of the labor institutions on Java grew out of labor surplus conditions, whereas for decades labor availability for rice production in Pinrang-Sidrap has been scarce relative to the demand for labor services. Family labor is used for most preharvest tasks in rice production in response to the unavailability or high cost of hired labor. *Gotong-royong* traditionally has been used for the labor-intensive task of planting to ensure labor during peak periods of demand.

Throughout the region, the institution of *gotong-royong* for planting is being replaced by hired labor. The employment of hired labor for planting began in the early 1980s in most villages, and it is still widely used. The surveyed farmers who use hired labor report that they changed their practice because of the unavailability of *gotong-royong* labor for planting on a timely basis. Farmers who crop technically irrigated *sawah* are particularly dependent on timely production practices because of their irrigation schedules. The *tebasan* system for harvesting has not been introduced in Pinrang-Sidrap. Harvesting throughout the region continues to be done by the *bawon* system, and the harvesters receive one-tenth of the harvest.

Implications of Institutional Change

Increased rice production, rural economic growth, and the expansion of off-farm employment opportunities for unskilled workers have led to widespread changes in traditional hiring and payment practices in all of the survey regions. On Java, the use of institutional arrangements designed to spread returns to labor among a large number of workers is declining as demand for labor from outside the rice sector rises. Market wages have become a more prominent means of payment and labor allocation. The asymmetric adjustment in wages that normally occurs in the process of economic development—inflexibility of wages to move below a subsistence level under labor surplus conditions but flexible upward movement in wages in response to competitive conditions—has been reinforced by changes in labor-hiring practices in the Indonesian rice economy.

New institutional practices governing the organization of labor in the rice sector give increasing attention to emerging labor scarcities. Contract labor groups have replaced daily wage labor in many regions, particularly for the labor-intensive tasks of planting, weeding, and harvesting. The use of contract labor reduces the possibility of labor shortages at peak seasons and enhances the efficiency of performance of each task. Groups are paid on a piecework basis according to the amount of land cultivated or the amount of paddy harvested. These adjustments in labor-hiring arrangements in response to rises in workers' marginal productivity are consistent with the development of an increasingly competitive labor market in rural Indonesia.

Wage Trends in the Rice Sector

The reduced prominence of labor surplus conditions in rural Java is reflected in the pattern of real wages within the rice sector as well as in changes in the use of traditional employment practices. This section examines secondary data on real wages for individual tasks in rice production from 1976 to 1988.[3] The analysis is confined to the main rice-growing provinces on Java (West, Central, and East Java), where unskilled labor moves relatively freely between the farm and nonfarm sectors of the rural economy and hired labor is used extensively in the production of rice.

The Choice of Deflators

Widespread disagreement exists in the literature and among current researchers in Indonesia about the path of real wages in the rice sector. The controversy rests on two issues—the choice of deflator applied to the nominal wages, and the region and time period of analysis. Mazumdar and Sawit (1986) and Naylor (1989) have shown that a wide range of real wage trends can be calculated from available price data in Indonesia, even if a single series of nominal wages is used for all calculations. The results vary from a stagnant trend in real wages between 1976 and 1988, when a comprehensive index of household expenditures is used to deflate nominal wages, to a 50 to 100 percent rise in real wages during the same period, when a simple rice price deflator is applied to nominal wages. This inconsistency in results makes it impossible accurately to assess real income growth and changes in the structure of the rural labor market based on credible real wage trends.

[3]Data on nominal wage rates have been collected regularly by the Central Bureau of Statistics (BPS) since 1976 in provinces on Java and since 1980 in ten other provinces. The BPS survey includes 180 farmers in West Java, 261 farmers in Central Java, 261 farmers in East Java, and 36 farmers in Yogyakarta. The number of farmers in the other provinces is much smaller. For further reference on these surveys, see Korns (1988) and Godfrey (1987). The analysis in this section focuses on hoeing (male) and weeding (female) wages from the BPS survey. These wages are paid largely in cash rather than in kind.

Income and price changes in the rural economy lead to systematic changes in consumption bundles over time, which bias estimates of real wage trends based on any deflator with fixed expenditure weights.[4] Although rice constitutes a major expenditure item of rural households (between 20 and 25 percent for low- and medium-income households), the choice of the rice price as a deflator does not incorporate the full range of household expenditures such as clothing, housing (purchase or building materials), and transportation. As incomes rise and a smaller proportion of household income is spent on food items, the use of a broader index of household expenditures becomes essential.[5]

Errors in weighting and calculating the various components of a comprehensive household expenditure index are more prevalent than in constructing a simple index of rice prices. Close examination of Indonesia's rural consumer price index by province reveals large distortions within the relatively minor expenditure category of chilies.[6] Removal of the misleading component of the index yields a plausible representation of real wages.[7]

The results obtained by using a revised index of consumption expenditures (excluding chilies) as a deflator for wages in rice production are shown in Figures 5.1–5.6.[8] Data for hoeing are used to represent move-

[4]The Laspeyres index and the Paasche index are commonly used price indexes in economic welfare analysis. The Laspeyres index weights prices by quantities in the base year, whereas the Paasche index weights prices by quantities in the final year. Accordingly, the Laspeyres index overestimates a rise in the cost of the original bundle and the Paasche index underestimates an increase in cost. The consumer price index paid by farmers that is used in this analysis is calculated by the Laspeyres weighting method. As a result, real wage trends based on the index tend to be understated. Because the results err on the conservative side, a calculated increase in real wages would indicate a definite improvement in the welfare of agricultural laborers.

[5]This trend is based on Engel's Law, which states that the income elasticity for food in the aggregate is less than one and approaches zero as incomes rise.

[6]The rural consumer price index used in this analysis is the household consumption component of the farmers' terms of trade index, 1976 = 100 (BPS). This index includes the costs of food, housing, clothing, and other goods and services, weighted by average per capita expenditures for rural households in each province. The consumer price data in the farmers' terms of trade index are often used as a deflator in the evaluation of real wages in rice production since both the wage and price series are collected simultaneously from the same sample.

[7]Extraordinary growth in chili prices between 1981 and 1988 caused the food component of the consumer price index to rise disproportionately. Further investigation of the reporting process at the village level is necessary before recorded chili prices can be used with confidence as part of a deflator of rural household expenditures. The BPS released a revised index of the farmers' terms of trade in mid-1989, which has a new base of 1983 = 100. Because of substantial changes in all expenditure categories and reweighting among the components of the index, the original and revised price series cannot be spliced to form a consistent deflator from 1976 to 1988. For further detail on the index problem, see Naylor (1990).

[8]The revised index consists of the index of prices paid by farmers in the BPS farmers' terms of trade series without the chili component.

FIGURE 5.1. *Real hoeing wage in West Java*

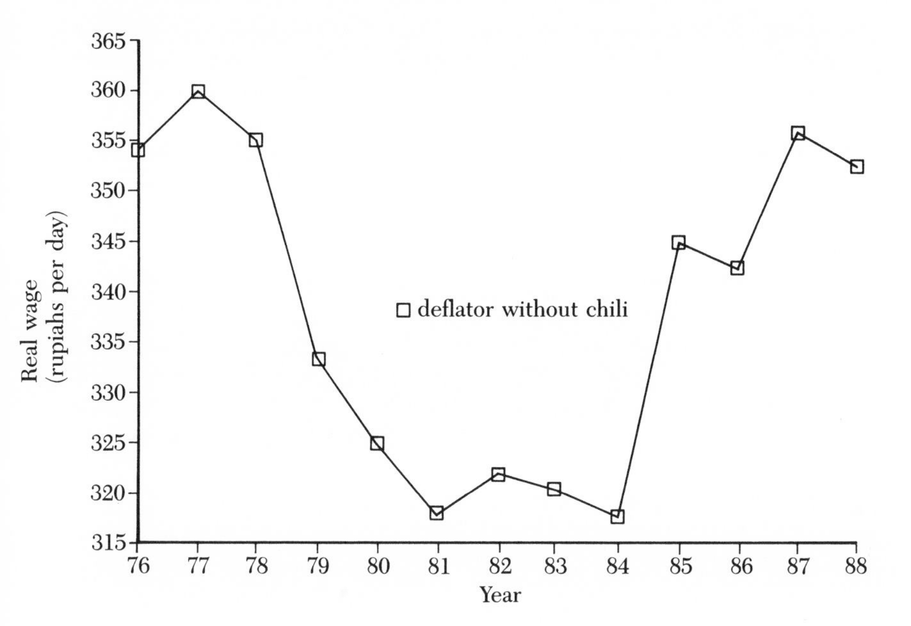

FIGURE 5.2. *Real weeding wage in West Java*

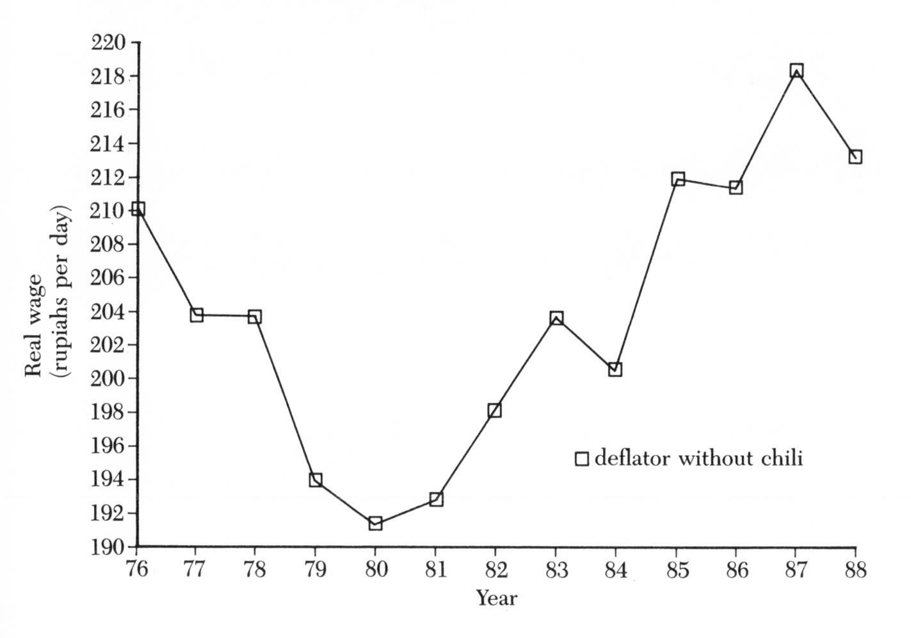

FIGURE 5.3. *Real hoeing wage in Central Java*

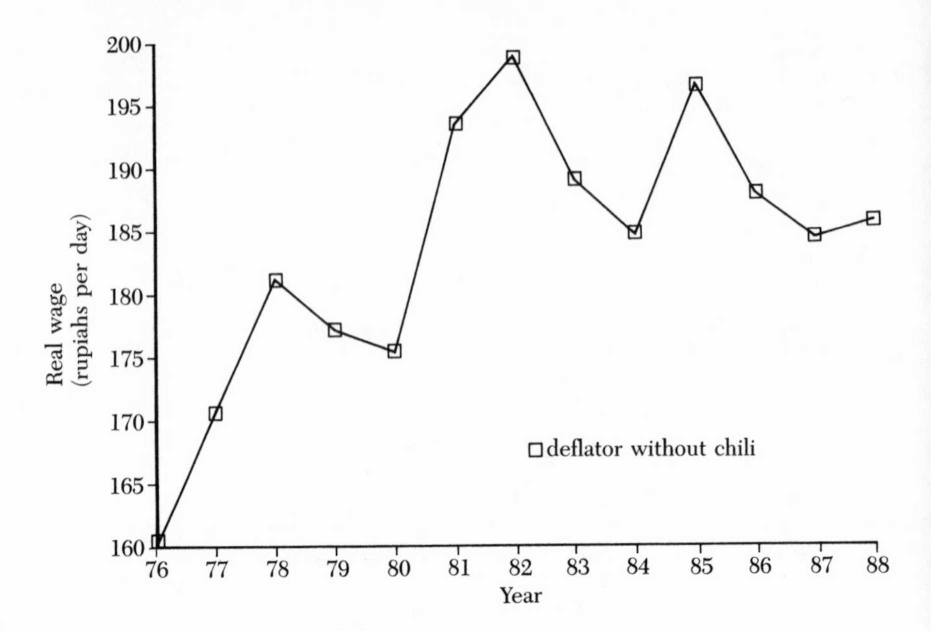

FIGURE 5.4. *Real weeding wage in Central Java*

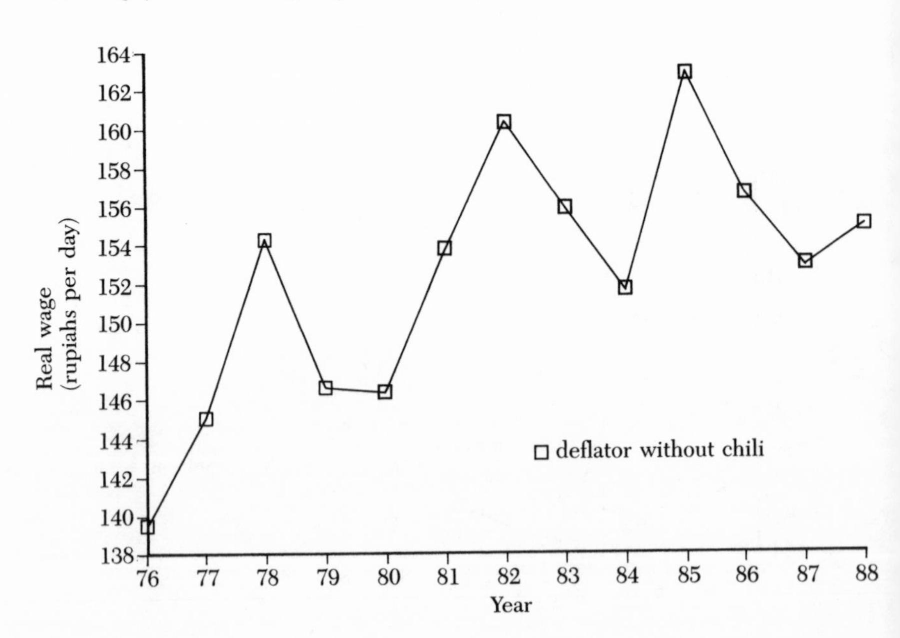

FIGURE 5.5. *Real hoeing wage in East Java*

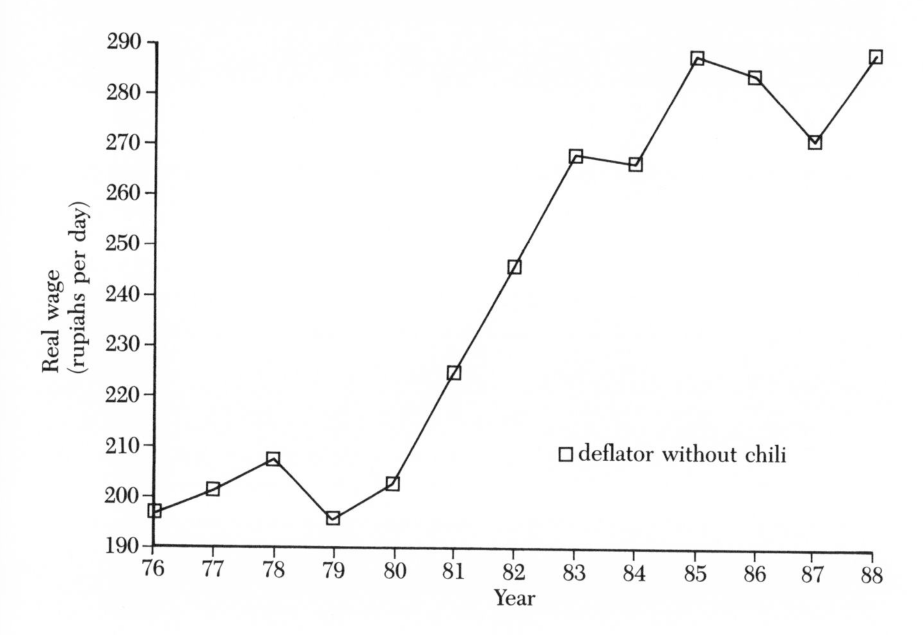

FIGURE 5.6. *Real weeding wage in East Java*

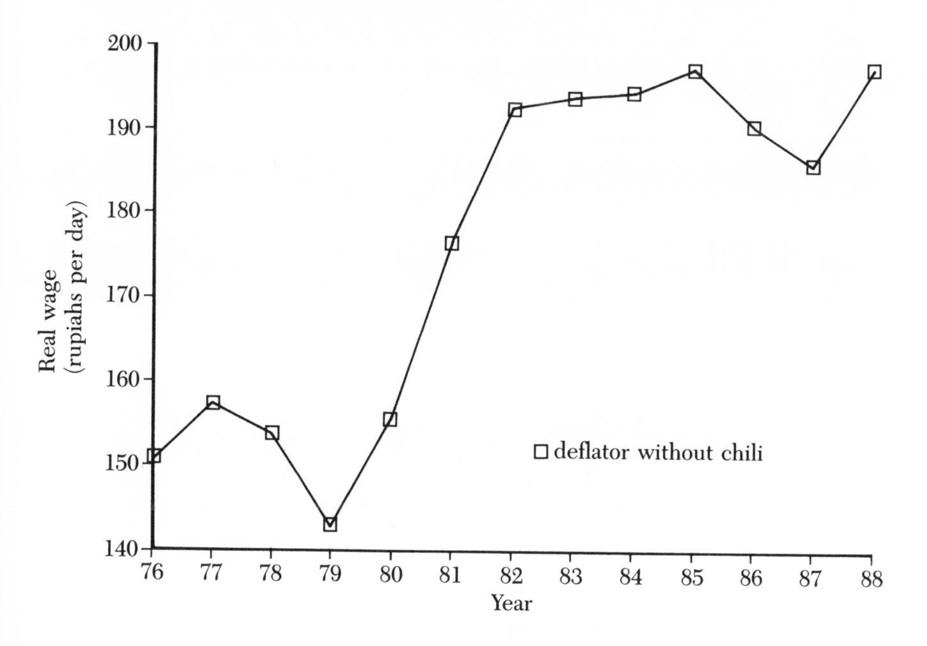

ments in male wages; those for weeding represent female wages. Trends in real wages rise throughout the first half of the 1980s for both male and female tasks in the main rice-producing provinces on Java. Growth in real wages during the early 1980s is consistent with several reinforcing influences in the Indonesian economy: rising demand for labor in off-farm activities stimulated by revenues earned from rice production; increased government spending in rural areas leading to the expansion of employment opportunities; and the economy-wide effects of the oil boom on the urban demand for labor (Manning 1986a, 1986b; Godfrey 1987; Collier et al. 1988). Although the increase in real wages does not show the cause of growing labor demand, wage movements are consistent with a tightening of the rural labor market and an absence of surplus labor.

Region and Time Period of Analysis

Figures 5.1–5.6 also illustrate the regional disparity in real wages between 1976 and 1988. Real wage growth was rapid in East Java, slower in Central Java, and almost stagnant in West Java. There is no consistent difference between the trends in male and female wages across regions. Wage levels for both male and female tasks remained higher in West Java than in East and Central Java throughout the reporting period (Table 5.7). Relatively high wage levels in West Java are the result of earlier rapid growth associated with expansion of the Jakarta economy. Nevertheless, the data show some convergence in wage levels. The fastest growth in wages during 1976–88 occurred in East and Central Java, the provinces with the lowest initial wage rates. A converging pattern of agricultural wages across regions reflects broad-based rural economic growth consistent with the expansion in the rice economy throughout Java during the early 1980s.

Despite their rapid growth in the 1980s, wages in rice production in Central Java remain substantially below those paid in East and West Java. The larger supply of labor in Central Java compared with that of other provinces on Java is probably the principal explanation for the relatively low wage levels. In 1985, Central Java had a reported population density of almost 800 people per square kilometer, whereas population densities in East and West Java were roughly 650 people per square kilometer (BPS 1987). Because transportation within and between provinces on Java is well developed, wage differentials most likely reflect the psychic and resource costs of migration.

Two recent disquieting turns of events in the rural labor market are apparent from the aggregate data. Between 1985 and 1987, real wages for unskilled workers in the rice sector remained stagnant and in some years

TABLE 5.7. *Nominal and real hoeing and weeding wages on Java, 1976–1988 (Rp per day)*

	WEST JAVA			
	Hoeing wages		Weeding wages	
Year	Nominal	Real[a]	Nominal	Real[a]
1976	354	354	210	210
1977	388	360	220	204
1978	408	355	234	204
1979	454	334	264	194
1980	532	325	313	191
1981	598	318	363	193
1982	668	322	411	198
1983	804	320	512	204
1984	916	318	579	201
1985	1,048	345	644	212
1986	1,149	343	709	211
1987	1,336	356	820	219
1988	1,483	353	897	213
	CENTRAL JAVA			
	Hoeing wages		Weeding wages	
Year	Nominal	Real[a]	Nominal	Real[a]
1976	160	160	140	140
1977	183	171	156	145
1978	206	181	175	154
1979	239	177	198	147
1980	287	176	239	146
1981	368	193	293	154
1982	418	199	337	160
1983	488	189	402	156
1984	563	185	462	152
1985	644	197	534	163
1986	710	188	592	157
1987	812	185	673	153
1988	856	186	714	155
	EAST JAVA			
	Hoeing wages		Weeding wages	
Year	Nominal	Real[a]	Nominal	Real[a]
1976	197	197	151	151
1977	219	201	171	157
1978	245	208	182	154
1979	285	196	208	143
1980	355	203	272	156
1981	456	225	358	177
1982	555	246	435	193
1983	759	268	549	194
1984	893	266	652	195
1985	1,021	288	700	198
1986	1,093	284	734	191
1987	1,191	271	816	186
1988	1,275	289	873	198

SOURCE: BPS, "Farmers' Terms of Trade Index (1978 = 100)."
[a]Real wages using rural consumer price index without chilies as a deflator (1976 = 100).

declined in East and Central Java (Table 5.7). In addition, the converging pattern in real wage trends across provinces that was evident during the early 1980s was reversed in recent years. Real wage growth in West Java, the province with the highest wage levels, exceeded that in East and Central Java between 1985 and 1987. Wage levels in Central Java fell further below those in East and West Java during the same period. Stagnant real wage rates in East and Central Java seem to indicate that improvements in labor welfare, which were pronounced during the first half of the 1980s, have not been sustained since 1985. Slower growth in labor demand in both urban and rural locations directly reflects the temporary downturn in economic activity associated with the oil-led recession in Indonesia.

More time and evidence are needed before definite conclusions regarding welfare changes can be made, however. A surge in economic growth, led by marked improvements in manufacturing output and nonoil exports, has begun as Indonesia enters the REPELITA V period (the fifth five-year plan, 1989–93). Real wages in rice production in East and Central Java increased in 1988 and could rise substantially during the next five years, especially if labor demand expands in nonagricultural sectors and in the production of agricultural commodities such as horticultural crops and livestock products, which are labor-intensive and have high income elasticities of demand.

Conclusion

Field survey results on employment opportunities for unskilled workers, adjustments in labor-hiring practices in the rice sector, and rising real wages in agricultural and nonagricultural activities provide evidence of a broadly competitive and well-integrated rural labor market on Java in the 1980s. Employment in the agricultural sector remains an important source of income for landless laborers on a seasonal basis. But nonfarm activities play an increasingly dominant role in contributing to rural household employment and income. Wages in rice cultivation are roughly comparable to wages for unskilled labor in seasonal jobs in rural nonagricultural activities. Unskilled workers face a wide range of income-earning alternatives and allocate their time to various activities depending on relative returns.

In response to increased competition in the labor market, many farmers on Java have discarded the use of traditional labor-hiring practices, previously designed to accommodate large surpluses of labor, and instead pay cash wages to ensure adequate and timely availability of labor. Contract arrangements are used extensively in rice production, especially in planting. The evolution of payment and employment practices reflects changing demand and supply in the labor market. Labor surplus conditions, defined by an unlimited supply of labor available at a minimum subsistence wage, no longer exist on Java. Wages for agricultural workers are determined endogenously by demand and supply in the rural labor market, and farmers pay competitive wages to secure labor on a timely basis.

Appendix 5.1. Field Survey Data on Wages in Rice Production

Appendix 5.1 contains data on local wages by task in rice production for the survey areas on Java. The appendix is designed to supplement the data provided in the text for readers interested in the level and diversity of wage rates at the local level. The first section of data includes wage rates (Rp per day and Rp per hour) for hoeing (male) and weeding (mostly female) tasks in rice production by village for each of the survey regions on Java. The regions are presented in the following order: Kabupaten Klaten in Central Java (Tables A5.1–5.4); Kabupaten Majalengka in West Java (Tables A5.5–5.8); and Kabupaten Kediri in East Java (Tables A5.9–5.12).

The data on hoeing and weeding wages by region are followed by village-level data on planting wages under contract arrangements for Kabupaten Klaten in Central Java (Tables A5.13–5.14) and Kabupaten Kediri in East Java (Table A5.15). Contract planting arrangements were not common in 1988 in Kabupaten Majalengka in West Java. The wage data for planting provide a cost comparison between the contract and noncontract systems of hiring labor for planting. The final table (A5.16) illustrates the wide variation in harvest shares paid to labor in the *bawon* and *tebasan* systems between villages in Kabupaten Klaten (Central Java).

TABLE A5.1. *Male wages for hoeing, Kabupaten Klaten (rupiahs per day, meals included)*

Village	1983	1987	1988	Hours per day
Kec. Kebonarum				
Desa Ngrundul	1,000	1,200	1,500	7
Desa Basin	750	1,000	1,500	7
Kec. Klaten Tengah				
Desa Jomboran	1,000	1,500	2,000	8
Desa Gumulan	750	1,250	1,500	7
Kec. Jogonalan				
Desa Bakung	750	1,000	1,250	8
Desa Rejoso	1,200	1,300	1,500	5
Kec. Karangnongko				
Desa Banyuaeng	1,000	1,500	1,500	7
Desa Demakijo	750	1,000	1,500	6
Kec. Weti				
Desa Pandes	750	1,100	1,200	5
Desa Canan	800	1,300	1,500	5
Kec. Ceper				
Desa Jumbu Kulon	500	750	850	4
Desa Pokak	1,000	1,500	1,500	6
Kec. Gantiwarno				
Desa Mlese	850	1,100	1,400	7
Desa Ngandong	1,000	1,250	1,500	8
Kec. Karang Anom				
Desa Padas	1,000	1,500	1,500	5
Desa Jeblog	750	1,000	1,250	6
Kec. Jatinom				
Desa Krajan	1,000	1,500	1,500	6
Kec. Polan-Harjo				
Desa Polan			1,500	5
Desa Jimur		1,500	1,500	6
Kec. Bayat				
Desa Tegalrejo			650	3
Desa Ngerangan			750	3
Kec. Cawes				
Desa Cawes	400	700	900	3
Desa Burikan			750	4
AVERAGE	847	1,208	1,326	6

TABLE A5.2. *Male wages for hoeing, Kabupaten Klaten (rupiahs per hour, meals included)*

Village	1983	1987	1988
Kec. Kebonarum			
Desa Ngrundul	143	171	214
Desa Basin	107	143	214
Kec. Klaten Tengah			
Desa Jomboran	125	188	250
Desa Gumulan	107	179	214
Kec. Jogonalan			
Desa Bakung	94	125	156
Desa Rejoso	240	260	300
Kec. Karangnongko			
Desa Banyuaeng	143	214	214
Desa Demakijo	125	167	250
Kec. Weti			
Desa Pandes	150	220	240
Desa Canan	160	260	300
Kec. Ceper			
Desa Jumbu Kulon	143	214	243
Desa Pokak	167	250	250
Kec. Gantiwarno			
Desa Mlese	121	157	200
Desa Ngandong	125	156	188
Kec. Karang Anom			
Desa Padas	200	300	300
Desa Jeblog	125	167	208
Kec. Jatinom			
Desa Krajan	167	250	250
Kec. Polan-Harjo			
Desa Polan			300
Desa Jimur		250	250
Kec. Bayat			
Desa Tegalrejo			217
Desa Ngerangan			250
Kec. Cawes			
Desa Cawes	133	233	300
Desa Burikan			188
AVERAGE	143	205	239

TABLE A5.3. *Female wages for weeding, Kabupaten Klaten (rupiahs per day, meals included)*

Village	1983	1987	1988	Hours per day
Kec. Kebonarum				
Desa Ngrundul	500	600	700	7
Desa Basin	750	1,000	1,500	7
Kec. Klaten Tengah				
Desa Jomboran	500	750	1,300	8
Desa Gumulan	300	600	800	7
Kec. Jogonalan				
Desa Bakung	600	800	1,000	8
Desa Rejoso	800	1,000	1,200	6
Kec. Karangnongko				
Desa Demakijo	500	750	1,000	8
Kec. Weti				
Desa Pandes	500	750	1,000	5
Desa Canan	750	1,000	1,200	5
Kec. Ceper				
Desa Jumbu Kulon	350	600	750	4
Desa Pokak	600	1,000	1,000	6
Kec. Gantiwarno				
Desa Mlese	500	700	900	7
Desa Ngandong	400	650	800	6
Kec. Karang Anom				
Desa Padas	200	600	600	5
Desa Jeblog	500	750	1,000	6
Kec. Polan-Harjo				
Desa Polan		850	950	5
Desa Jimur	500	800	1,050	6
Kec. Bayat				
Desa Tegalrejo			450	3
Desa Negerangan			500	3
Kec. Cawes				
Desa Cawes	300	400	500	3
Desa Burikan			500	4
AVERAGE	503	756	890	6

TABLE A5.4. *Female wages for weeding, Kabupaten Klaten (rupiahs per hour, meals included)*

Village	1983	1987	1988
Kec. Kebonarum			
Desa Ngrundul	71	86	100
Desa Basin	107	143	214
Kec. Klaten Tengah			
Desa Jomboran	63	94	163
Desa Gumulan	43	86	114
Kec. Jogonalan			
Desa Bakung	75	100	125
Desa Rejoso	133	167	200
Kec. Karangnongko			
Desa Demakijo	63	94	125
Kec. Weti			
Desa Pandes	100	150	200
Desa Canan	150	200	240
Kec. Ceper			
Desa Jumbu Kulon	88	150	188
Desa Pokak	100	167	167
Kec. Gantiwarno			
Desa Mlese	71	100	129
Desa Ngandong	67	108	133
Kec. Karang Anom			
Desa Padas	40	120	120
Desa Jeblog	83	125	167
Kec. Polan-Harjo			
Desa Polan		170	190
Desa Jimur	83	133	175
Kec. Bayat			
Desa Tegalrejo			150
Dewsa Ngerangan			167
Kec. Cawes			
Desa Cawes	100	133	167
Desa Burikan			125
AVERAGE	85	129	160

TABLE A5.5. *Male wages for hoeing, Kabupaten Majalengka (rupiahs per day, meals included)*

Village	1983	1987	1988	Hours per day
Kec. Leuwianghindang				
Desa Weragati	1,250		2,500	6
Desa Pasir			2,500	5
Kec. Rajalgaluh				
Desa Rjl. Kidul			2,000	8
Desa Sungawangi	1,500	2,000	2,500	8
Kec. Majalengka				
Desa Sidamukti	1,500	2,000	2,500	8
Kec. Cigasong				
Desa Simpureum	1,000	2,500	3,000	8
Kec. Jatiwangi				
Desa Leuweunjede	1,500	2,000	2,500	8
Desa Cibentar		3,000	3,000	8
Kec. Dawuan				
Desa Kasokandel	1,000	1,500	2,000	8
Desa Genteng	1,000	1,500	1,750	7
Kec. Kertajati				
Desa Sukawana	1,000	1,500	1,500	6
Desa Kertaninagin	1,000	1,600	2,000	6
Kec. Ligung				
Desa Sukawera		1,500	2,000	8
Desa Leuweu.	1,500	2,250	3,000	12
Kec. Maja				
Desa Paniis	1,000	1,250	1,500	5
Desa Pageraji	1,000	1,250	1,500	5
Kec. Bantarujeg				
Desa Bantarujeg	2,000	2,000	3,000	8
Desa Babakansari	1,000	1,750	2,500	8
Kec. Cikijing				
Desa Banjaransari	1,200	2,500	2,500	6
Desa Sindang	1,500	2,000	2,500	5
AVERAGE	1,247	1,888	2,313	7

TABLE A5.6. *Male wages for hoeing, Kabupaten Majalengka (rupiahs per hour, meals included)*

Village	1983	1987	1988
Kec. Leuwianghindang			
Desa Weragati	208		417
Desa Pasir			500
Kec. Rajalgaluh			
Desa Rjl. Kidul			250
Desa Sungawangi	188	250	313
Kec. Majalengka			
Desa Sidamukti	188	250	313
Kec. Cigasong			
Desa Simpureum	125	313	375
Kec. Jatiwangi			
Desa Leuweunjede	188	250	313
Desa Cibentar		375	375
Kec. Dawuan			
Desa Kasokandel	125	188	250
Desa Genteng	143	214	250
Kec. Kertajati			
Desa Sukawana	167	250	250
Desa Kertaninagin	167	267	333
Kec. Ligung			
Desa Sukawera		188	250
Desa Leuweu.	125	188	250
Kec. Maja			
Desa Paniis	200	250	300
Desa Pageraji	200	250	300
Kec. Bantarujeg			
Desa Bantarujeg	250	250	375
Desa Babakansari	125	219	313
Kec. Cikijing			
Desa Banjaransari	200	417	417
Desa Sidang	300	400	500
AVERAGE	181	266	332

TABLE A5.7. *Female wages for weeding, Kabupaten Majalengka (rupiahs per day, meals included)*

Village	1983	1987	1988	Hours per day
Kec. Leuwianghindang				
Desa Weragati			1,000	6
Desa Pasir			1,100	5
Kec. Rajalgaluh				
Desa Rjl. Kidul			1,000	8
Desa Sungawangi		1,000	1,500	8
Kec. Majalengka				
Desa Sidamukti	1,000	1,500	1,700	8
Kec. Cigasong				
Desa Simpureum	500	1,000	1,500	5
Kec. Jatiwangi				
Desa Leuweunjede	250	500	1,000	5
Desa Cibentar		1,000	1,000	6
Kec. Dawuan				
Desa Kasokandel	500	800	1,000	7
Desa Genteng	500	750	1,000	7
Kec. Kertajati				
Desa Sukawana	500	750	750	6
Desa Kertaninagin	400	750	750	6
Kec. Ligung				
Desa Sukawera	500	750	1,000	5
Desa Leuweu.	700	1,000	1,000	6
Kec. Maja				
Desa Paniis	400	500	700	5
Desa Pageraji	500	600	750	5
Kec. Bantarujeg				
Desa Bantarujeg	1,000	1,000	1,500	5
Desa Babakansari	500	750	1,000	5
Kec. Cikijing				
Desa Banjaransari	500	1,250	1,250	6
Desa Sindang	600	750	1,000	5
AVERAGE	557	862	1,075	6

TABLE A5.8. *Female wages for weeding, Kabupaten Majalengka (rupiahs per hour, meals included)*

Village	1983	1987	1988
Kec. Leuwianghindang			
Desa Weragati			167
Desa Pasir			220
Kec. Rajalgaluh			
Desa Rjl. Kidul			125
Desa Sungawangi		125	188
Kec. Majalengka			
Desa Sidamukti	125	188	213
Kec. Cigasong			
Desa Simpureum	100	200	300
Kec. Jatiwangi			
Desa Leuweunjede	50	100	200
Desa Cibentar		167	167
Kec. Dawuan			
Desa Kasokandel	71	114	143
Desa Genteng	71	107	143
Kec. Kertajati			
Desa Sukawana	83	125	125
Desa Kertaninagin	67	125	125
Kec. Ligung			
Desa Sukawera	100	150	200
Desa Leuweu.	117	167	167
Kec. Maja			
Desa Paniis	80	100	140
Desa Pageraji	100	120	150
Kec. Bantarujeg			
Desa Bantarujeg	200	200	300
Desa Babakansari	100	150	200
Kec. Cikijing			
Desa Banjaransari	83	208	208
Desa Sindang	120	150	200
AVERAGE	98	147	184

TABLE A5.9. *Male wages for hoeing, Kabupaten Kediri (rupiahs per day, meals included)*

Village	1982	1987	Hours per day
Kec. Kediri			
Kota Kediri	1,000	1,750	6
Kec. Mojo			
Desa Jugo	600	900	5
Desa Purowasri	750	1,250	6
Kec. Gurah			
Desa Gempolan	750	1,250	7
Desa Tampkrejo	850	1,250	6
Kec. Pagu			
Desa Bulupasar	750	1,400	5
Kec. Gampengrejo			
Desa Gorgorante	800	1,500	7
Kec. Wates			
Desa Tempurejo	400	650	3
Kec. Pare			
Desa Palem	400	700	4
Desa Tulungrejo	900	1,500	6
Kec. Kandangan			
Desa Kandangan	700	1,000	4
Desa Kalampison	900	1,250	8
Kec. Ngadiluwih			
Desa Ngatiluruh	450	650	3
Kec. Plotoklaten			
Desa Klanderan	500	700	4
AVERAGE	696	1,125	5

TABLE A5.10. *Male wages for hoeing, Kabupaten Kediri (rupiahs per hour)*

Village	1982	1987
Kec. Kediri		
Kota Kediri	167	292
Kec. Mojo		
Desa Jugo	120	180
Desa Purowasri	125	208
Kec. Gurah		
Desa Gempolan	107	179
Desa Tampkrejo	142	208
Kec. Pagu		
Desa Bulupasar	150	280
Kec. Gampengrejo		
Desa Gorgorante	114	214
Kec. Wates		
Desa Tempurejo	133	217
Kec. Pare		
Desa Palem	100	175
Desa Tulungrejo	150	250
Kec. Kandangan		
Desa Kandangan	175	250
Desa Kalampison	113	156
Kec. Ngadiluwih		
Desa Ngatiluruh	150	217
Kec. Plotoklaten		
AVERAGE	134	217

TABLE A5.11. *Female wages for weeding, Kabupaten Kediri (rupiahs per day, meals included)*

Village	1982	1987	Hours per day
Kec. Mojo			
Desa Jugo	250	600	5
Kec. Purowasri			
Desa Wonotengah	600	1,000	6
Kec. Gurah			
Desa Gempolan	500	800	5
Kec. Pagu			
Desa Bulupasar	400	850	5
Kec. Gampengrejo			
Desa Gorgorante	350	600	4
Kec. Wates			
Desa Tempurejo	200	300	3
Kec. Pare			
Desa Palem	350	550	4
Desa Tulungrejo	700	900	6
Kec. Kandangan			
Desa Kandangan	500	750	4
Desa Kalampison	500	750	4
Kec. Ngatiluruh			
Desa Ngatiluruh	300	550	3
Kec. Plotoklaten			
Desa Klanderan	300	550	4
AVERAGE	413	683	4

TABLE A5.12. *Female wages for weeding, Kabupaten Kediri (rupiahs per hour, meals included)*

Village	1982	1987
Kec. Mojo		
Desa Jugo	50	120
Kec. Purowasri		
Desa Wonotengah	100	167
Kec. Gurah		
Desa Gempolan	100	160
Kec. Pagu		
Desa Bulupasar	80	170
Kec. Gampengrejo		
Desa Gorgorante	88	150
Kec. Wates		
Desa Tempurejo	67	100
Kec. Pare		
Desa Palem	88	138
Desa Tulungrejo	117	150
Kec. Kandangan		
Desa Kandangan	125	188
Desa Kalampison	125	188
Kec. Ngatiluruh		
Desa Ngatiluruh	100	183
Kec. Plotoklaten		
Desa Klanderan	75	138
AVERAGE	93	154

TABLE A5.13. *Female contract wages for planting, Kabupaten Klaten (rupiahs per day, meals included)*

Village	1983	1987	1988	Hours per day	Number of workers
Kec. Kebonarum					
Desa Ngrundul	22,000	26,000	30,000	8	24
Desa Basin	18,000	20,000	24,000	8	24
Kec. Klaten Tengah					
Desa Jomboran	17,000	30,000	35,000	4	30
Desa Gumulan	12,000	20,000	28,000	10	24
Kec. Jogonalan					
Desa Bakung	28,000	32,000	36,000	4	32
Desa Rejoso	25,000	30,000	40,000	4	40
Kec. Karangnongko					
Desa Banyuaeng	12,000	16,000	24,000	8	16
Desa Demakijo	18,000	24,000	30,000	6	30
Kec. Weti					
Desa Pandes	25,000	30,000	35,000	6	50
Desa Canan	22,000	27,000	30,000	5	25
Kec. Ceper					
Desa Jumbu Kulon	11,250	22,500	26,000	6	18
Desa Pokak	25,000	30,000	30,000	6	40
Kec. Gantiwarno					
Desa Mlese	25,000	32,000	40,000	7	50
Kec. Karang Anom					
Desa Padas	19,200	24,000	24,000	8	16
Desa Jeblog	15,000	20,000	25,000	8	30
Kec. Jatinom					
Desa Krajan		27,000	32,800	12	20
Kec. Polan-Harjo					
Desa Polan			20,000	6	25
Desa Jimur	20,000	32,000	38,000	4	40
AVERAGE	19,653	26,029	30,433	7	30

TABLE A5.14. *Female contract wages for planting, Kabupaten Klaten (rupiahs per hour, meals included)*

Village	1983	1987	1988
Kec. Kebonarum			
Desa Ngrundul	115	135	156
Desa Basin	94	104	125
Kec. Klaten Tengah			
Desa Jomboran	142	250	292
Desa Gumulan	50	83	117
Kec. Jogonalan			
Desa Bakung	219	250	281
Desa Rejoso	156	188	250
Kec. Karangnongko			
Desa Banyuaeng	94	125	188
Desa Demakijo	100	133	167
Kec. Weti			
Desa Pandes	83	100	117
Desa Canan	176	216	240
Kec. Ceper			
Desa Jumbu Kulon	104	208	241
Desa Pokak	104	125	125
Kec. Gantiwarno			
Desa Mlese	71	91	114
Kec. Karang Anom			
Desa Padas	150	188	188
Desa Jeblog	63	83	104
Kec. Jatinom			
Desa Krajan		113	137
Kec. Polan-Harjo			
Desa Polan			133
Desa Jimur	125	200	238
AVERAGE	115	153	178

TABLE A5.15. *Female wages (contract or individual) for planting, Kabupaten Kediri, 1987 (meals included)*

Village	Contract cost (rupiahs)	Hours per contract	Number of workers	Wage per worker (rupiahs per hour)
Kec. Kediri	35,000	12	10	292
Kec. Pare	35,000	5	35	200
Kec. Kandangan	30,400	32	4	238
Kec. Ngadiluwih	33,500	9	10	372
Kec. Klanderan	35,000	7	15	333

Village	Wage (rupiahs per hour)
Kec. Mojo	120
Kec. Purwoasri	167
Kec. Gurah	250
Kec. Pagu	170

TABLE A5.16. *Harvest shares, Kabupaten Klaten, 1987–88*

Village	*Bawon*	*Tebasan*
Kec. Kebonarum		
Desa Ngrundul	1/10	1/20
Desa Basin	1/14	1/18
Kec. Klaten Tengah		
Desa Jomboran	1/10	1/15
Desa Gumulan	1/10	1/12
Kec. Jogonalan		
Desa Bakung	1/10	1/16
Desa Rejoso	1/15	1/20
Kec. Karangnongko		
Desa Banyuaeng	1/20	1/22
Desa Demakijo	1/15	1/25
Kec. Weti		
Desa Pandes	1/15	1/20
Desa Canan	1/15	1/20
Kec. Cepera[a]		
Desa Jumbu Kulon	1/20	1/25
Desa Pokak	1/15	1/20
Kec. Gantiwarno		
Desa Mlese	1/12	1/15
Desa Ngandong	1/10 (*ani-ani*) 1/12 (sickle)	1/15
Kec. Karang Anom		
Desa Padas	1/18	1/18
Desa Jeblog	1/17	1/17
Kec. Jatinom		
Desa Krajan	1/20	
Kec. Polan-Harjo		
Desa Polan	1/20	

[a]Wages paid for harvest by farmers.

6. Technical Change in Wetland Rice Agriculture

Paul Heytens

Indonesia's wetland rice farmers use a wide spectrum of production technologies. A variety of fertilizers and pesticides are available at subsidized prices; four different land preparation technologies are practiced throughout much of Indonesia; new seed varieties appear almost yearly; and several institutional arrangements for planting and harvesting coexist even in the same village. Farmers' selection of inputs and levels of use differ across the rice-producing regions, and these decisions affect productivity and the demand for labor.

The purposes of this chapter are threefold. The first goal is to investigate the dynamics of technical change since the inception of the Green Revolution and to describe the changes that have occurred in input use in wetland rice agriculture. A second aim is to examine how technical change has affected employment and farm incomes from rice production on Java. Answers to these questions indicate the extent to which the expansion of rice production since the late 1960s has promoted rural employment and income growth directly, important criteria in assessing alternative future rice strategies. A final task is to discuss the factors likely to affect the adoption of labor-displacing technology in the future.

Current Input Use in Indonesian Rice Production

The most striking observation from the farmer surveys is the ubiquity of the Green Revolution package of inputs. High-yielding seed varieties have been disseminated rapidly since the first improved seeds from IRRI were introduced to Indonesia in the late 1960s. Within the wetland systems of Java, adoption of HYVs is virtually complete. At present, the adoption rate is over 90 percent in the wetlands and about 83 percent for all rice land. This widespread replacement of traditional seed varieties with short-duration HYVs has transformed the nature of wetland rice agriculture in Indonesia from one of low yields, nonuse of purchased inputs, and single annual rice crops to one of high yields, high levels of purchased inputs, and multiple rice crops.

Another vivid impression from the field surveys is the regional variation in wage rates, labor-hiring arrangements, and degree of mechanization, both within Java and between on- and off-Java areas. The patterns of input use and institutional arrangements in wetland rice cultivation are spread like a mosaic across the Indonesian landscape, determined by relative wages, technical constraints, and the level of rice profits. This diversity of input use is discussed below in connection with each individual task in rice cultivation.

Land Preparation

In Indonesia, four individual technologies—hand hoes, draft animals, two-wheel tractors, and four-wheel tractors—are available for land preparation. Because of topographic and plot-size constraints, mechanical problems, and differing factor endowments of farmers, many technologies coexist in the same village. Very small plots tend to be hand-hoed. Farmers who own their own draft animals are likely to use them on their rice fields. Farmers who do not own animals generally use whichever technology is most cost-effective. The presence of large rocks or steep slopes sometimes prevents the use of tractors and draft animals. Finally, in many areas of Java, hand hoes are still used alongside animal traction or tractors, particularly to repair bunds and to turn corners difficult to reach with animals or machines.

On a regional scale, choice of technique appears to depend primarily on the relative prices of inputs. Of the three survey areas in Java, only land preparation in Majalengka, West Java, can be classified as partially mechanized. Hourly wages for hoers and animal power in Majalengka are higher than in either Klaten, Central Java, or Kediri, East Java. Consequently, the prevailing rate per hectare for plowing and harrowing by a two-wheel tractor (including some hoeing) in 1987 is competitive with other techniques in Majalengka, but low wages make tractor rental a less profitable venture in the other areas of Java. Hourly real wages for animal traction and hoeing in Kediri and Klaten would have to rise to levels prevailing in Majalengka before tractors would become an attractive alternative. The findings in the field mirror the situation in Java on a larger scale. Only West Java is tractorized to any significant degree.

In both of the off-Java fieldsites, the mechanization of land preparation is more advanced. Wages in Agam, West Sumatra, and Pinrang-Sidrap, South Sulawesi, are much higher than in any of the Java sites. Wages are higher because of shortages of manual laborers and draft animals at peak demand times such as during land preparation. In both areas, population density per hectare of *sawah* is much lower than in any of the Java fieldsites.

Generally, the level of tractorization in Indonesia is very low. Since the mid-1970s, the total number of tractors in Java and the outer islands has increased severalfold, but the actual percentage of land cultivated by machine is still very small. For example, the number of two-wheel tractors in Java increased from 3,996 in 1981 to 9,496 in 1986 (BPS 1987). But if one hand tractor cultivated as much as thirty-five hectares per year, total tractorized area would not have exceeded 7 percent of total cultivated wetland area in 1986. The corresponding figures in West Sumatra and South Sulawesi for 1986 are 3.5 and 10 percent if mini- and small, four-wheel tractors are included at a capacity of seventy hectares per year. (On Java, only two-wheel tractors are used on rice land, but off Java the smaller four-wheel tractors are also used.)

Many explanations underlie the low level of tractorization in Indonesia. There are no agronomic reasons a priori (Binswanger 1978) and no evidence empirically in Indonesia (Lingard and Bagyo 1983; Maamum et al. 1983) that tractor preparation confers any yield advantage over other land-preparation techniques. On Java, the generally low level of wages and technical factors, especially the small size of plots and the impracticality of using tractors in hilly areas, are the main constraints on mechanization of land preparation. The constraints on greater tractor use for the off-Java areas are probably more varied, although topographic limitations and the greater difficulty in obtaining and servicing tractors in the outer islands are important.

Government policy recently has not promoted tractorization. Although many owners purchased tractors at government-subsidized nominal interest rates of 12 percent (real rates of 2 to 4 percent) per annum since the mid-1970s, these concessionary rates are no longer available. Nominal borrowing rates have been over 20 percent per year for commercial loans during much of the 1980s. Moreover, tractors are not cheap in Indonesia. High costs of assembly (because of tariffs on imported parts and high-priced domestic parts) and large distribution costs (because of supply monopolies) contribute to high prices. Domestic prices of two- and four-wheel tractors were estimated to have been roughly 25 percent above world prices in 1988.

Planting

The transplanting of seedlings is practiced universally in the wetland rice systems of Indonesia. Seedlings are transplanted in rows by women on the main plot three to six weeks after the seeds have been planted in a nursery. This practice predated the dissemination of high-yielding variety seeds and other inputs. In this sense there has been no recent technical change in planting techniques. Minor technical innovations, such as using a small board to lay out grid marks for plant spacing and placing seedlings in straight rows, have made the task easier and faster, but the fundamental task remains the same. More interesting planting issues are the changing institutional and labor-hiring arrangements that farmers employ to obtain labor, discussed in Chapter 5.

Weeding

Weed growth is a problem in all wetland rice systems. The severity varies depending on the degree of water control. Weed problems tend to be greater when *sawah* alternately becomes wet and dry, a common phenomenon in rainfed and lower-productivity *sawah*, where water control is uneven. But weeds are present on all *sawah*. There are two basic approaches to dealing with weed problems. Farmers can either try to prevent weed growth or remove weeds after they appear. Rice farmers in Indonesia overwhelmingly take the second approach.

With the advent of straight-row transplanting of rice seedlings, it has become possible to use hand tools for weed removal during the early stages of rice plant growth. Farmers have used various homemade hand tools, including a rotary weeder that is pushed along the top of the soil. Although no evidence was available from the survey, other researchers have reported that the introduction of this practice halved labor use in the early weeding (Sinaga 1978; Collier et al. 1982a). In the later stages of the cropping season, the increase in size and breadth of the rice plant prevents the use of all implements to remove weeds. Therefore, weeding later in the season must be done by hand.

In Pinrang-Sidrap, farmers spray herbicides to reduce weed growth on higher-productivity, irrigated plots. Use of herbicides dramatically reduces or completely eliminates weed growth, thus allowing for large reductions in labor requirements. No herbicides were used in any of the Java fieldsites or in Agam, West Sumatra. In Java, the reasons probably stem from the low costs of labor and the perception that chemicals are ineffective in reducing weed growth. Also, herbicide use has not been promoted in government intensification programs in Java as it has in South Sulawesi. In Agam the reasons are less clear but probably stem from a lack of knowledge and availability; the higher labor costs there could make herbicide use an attractive weed-control option.

Fertilizer Use

Fertilizer subsidies have been one of the cornerstones of Indonesia's rice development program, and use per hectare is high in comparison with other rice-producing countries in Southeast Asia. The application of fertilizer has risen dramatically since the late 1960s. Nutrient sources have become more diversified in recent years, as indicated in Tables 6.1 and 6.2. Urea makes up a large but declining portion of total use; triple superphosphate (TSP) accounts for a good portion of the remainder. Fertilizer use is much higher on Java than in the outer islands, averaging 276

TABLE 6.1. *Fertilizer use per hectare on rice, 1969–86 (kilograms per hectare)*

Year	Total fertilizer	Urea
1969	31.1	26.9
1970	15.7	14.8
1971	38.7	34.1
1972	46.2	42.6
1973	66.9	55.7
1974	65.6	49.7
1975	74.4	55.2
1976	73.0	55.7
1977	95.2	77.0
1978	97.3	76.5
1979	114.9	91.2
1980	169.5	130.5
1981	205.0	148.6
1982	254.5	167.6
1983	258.7	161.8
1984	292.1	181.5
1985	305.4	180.5
1986	316.1	183.1

SOURCE: BPS.
NOTE: Assumes 70 percent of foodcrop fertilizer used on rice.

kilograms of urea, 99 kilograms of TSP, and 3 kilograms of other fertilizer per hectare on Java's wetlands and 106 kilograms of urea, 50 kilograms of TSP, and 6 kilograms of other fertilizers per hectare in wetland areas off Java (BPS, 1986). Fertilizer use in the Java survey areas was somewhat higher and more diversified than these overall averages (see Appendix 4.1, Chapter 4). Starting in 1987, farmers with higher-productivity, irrigated *sawah* in Kediri, Klaten, and Majalengka started using potassium chloride (KCl) and ammonium sulfate (ZA) as well as increased levels of TSP as part of the INSUS Packet D (general intensification) and SUPRA

TABLE 6.2. *Diversification of fertilizer sources, 1984–90 (000 metric tons)*

Fertilizer	1984–85	1985–86	1986–87	1987–88	1988–89	1989–90[a]
Urea	2,573	2,611	2,721	2,746	3,037	3,250
Ammonium sulfate (ZA)	417	465	511	561	648	680
Triple superphosphate (TSP)	994	1,087	1,210	1,209	1,426	1,600
Potassium chloride (KCl)	238	297	272	409	547	669

SOURCE: Ministry of Agriculture, BIMAS.
[a]Projected.

INSUS (super intensification) programs. For example, SUPRA INSUS recommendations for the 1989 wet season are that Java participants apply a standard package consisting of 250 kilograms of urea, 100 kilograms of TSP, 75 to 100 kilograms of KCl, and 100 kilograms of ZA (depending on the sulfur content of the local soils).

Fertilizer use at the off-Java fieldsites was lower than at the Java sites but somewhat higher than the overall average for the outer islands. Survey farmers off Java tended to use only urea and TSP. The more recent intensification programs recommending use of ZA and KCl had not begun in off-Java areas at the time of the field surveys.

The yield advantages from applying chemical fertilizer were clear to farmers participating in the field surveys. All stated that their yields had risen in response to higher fertilizer applications. Among survey farmers, fertilizer use was greater on the higher-productivity systems with good water control; fertilizer applied in a more stable and fertile crop environment was considered more likely to pay off and less risky than fertilizer applied in a variable environment. In the good-control *sawah* systems, farmers tended to apply less fertilizer during the wet season to reduce the risk of lodging. Lodging typically is not a problem during the dry seasons.

In virtually all of Indonesia's wetlands, fertilizer is broadcasted onto the rice paddy. Fertilizer generally is applied three times in a season—at the time of transplanting and twenty to thirty days and thirty to forty-five days afterward. In all survey areas, farmers used family labor to spread fertilizer. An average fertilizer application per hectare can easily be finished in a day by two or three people. Hence family labor usually is sufficient, particularly on the small *sawah* plots that characterize wetland rice production on Java.

Harvesting

The most commonly described technological change in Indonesian rice agriculture is the replacement of the traditional *ani-ani* (hand knife) with the sickle to harvest rice. As noted in Chapter 5, this shift involves considerably more than a change in technologies. The *ani-ani* has been associated in the literature with Geertz's notion of involution and the traditional *bawon* or share harvest open to all who wish to participate. The sickle, on the other hand, has been associated with the closing of the harvest and the *tebasan* system in which farmers sell their standing rice crop for cash to a middleman who conducts the harvest. In the early 1970s, the shift to sickle harvesting tended to involve a substantial change in social power relations (Collier et al. 1974; Hayami and Hafid 1979). As time passed and high-yielding seed varieties spread, adoption occurred for efficiency and technical reasons. By 1983, the sickle was the major harvesting tool in wet rice cultivation and was used by 70 to 80 percent of farmers (BPS, 1983). A more recent estimate for the wet season 1986–87 placed sickle use at 85 percent of all rice farmers in fifteen major rice-growing provinces.

The widespread adoption of the sickle along with HYVs makes sense for technical and economic reasons. The sickle is the most convenient and efficient tool to harvest the shorter and thicker-stemmed HYVs, which tend to lodge and ripen at the same time. With an *ani-ani* the harvester cuts each individual panicle, a task that is much easier to perform on the taller, traditional seed varieties. The increased cropping intensity made possible by the shorter-maturing HYVs made speed more essential, again greatly favoring the use of the sickle. Finally, because of the sickle's greater speed, farmers need to hire far fewer people to harvest a hectare of rice with a sickle than with the *ani-ani*, making harvests much easier to arrange in times when labor supplies are tight.

In the future, the major input-substitution issue in harvesting likely will be with threshing. Manual threshing (e.g., hitting paddy stalks on the ground or stomping by foot) is still by far the most commonly used method. But in the 1980s the number of fuel- and pedal-driven threshers in Indonesia rapidly increased. On Java, the number of machine threshers (both manual and fuel-driven) increased from 13,670 in 1981 to 54,313 in 1986. The corresponding numbers for West Sumatra were 176 in 1981 and 1,136 in 1986 and for South Sulawesi, 91 and 15,089, respectively.

These increases are rather dramatic, but they do not represent a very large portion of cultivated area (see below). Threshing by machine does not necessarily confer a cost advantage over manual techniques. For example, in Agam both the thresher owner and traditional harvesters received about a 10 percent share of the harvest, and the farmer had to pay extra to have the *padi* cut before machine threshing. Returns with machines are often higher because of reduced harvesting losses (estimated to be 5 to 10 percent of total output) and a higher-quality output—there are fewer brokens, drying is easier, and winnowing is more effective. In addition, machine threshing is much faster and less labor-intensive, a big advantage in the labor-scarce areas off Java.

The Impact of Technical Change on Farm Labor Demand and Income in Rice Production

Technical change can influence labor demand and farm income in two ways. It can do so directly by changing the labor-intensity and yield per hectare and also through intensification of crop production—raising cropping intensities on existing land. Change in the labor-intensity for a particular task usually implies a change in the number of workers employed in the agricultural sector. Raising cropping intensities increases the frequency of employment opportunities during the year and thus increases the total employment per worker.

Technical Change and Labor Demand

As shown in Table 6.3, total cultivated wetland rice area in Indonesia increased by 35 percent between 1969 and 1987. The corresponding figures for Java and off Java are 24 and 52 percent, respectively (the off-Java figure includes a large increase in low-productivity, tidal swamp area in the 1980s). On Java, the rise resulted mainly from an increase in cropping intensity; physical *sawah* area increased very slightly. Additional crops were made possible by adoption of the much shorter-duration HYV seeds (crop duration was cut from 6 to 3.5 months on average), which could be exploited as a result of public investments in new irrigation systems and rehabilitation.

Considerable field evidence on labor use by task in wetland rice agriculture in Java has been gathered on a regular basis by research teams since the late 1960s. Collier and his colleagues, in particular, have been active in the collection of such data. The various studies are far-ranging both in regional scope and time period and conclude, generally, that labor use per hectare per crop on Java has declined since the initial spread of HYVs.

Early studies show little difference in labor-intensity for preharvest

TABLE 6.3. *Cultivated wetland rice area, 1969–87 (000 hectares)*

Year	Java	Off Java	Total
1969	3,939	2,584	6,523
1970	3,958	2,706	6,664
1971	4,041	2,828	6,869
1972	4,010	2,625	6,635
1973	4,210	2,802	7,012
1974	4,426	2,883	7,309
1975	4,393	2,942	7,335
1976	4,233	3,008	7,241
1977	4,144	3,061	7,205
1978	4,430	3,213	7,643
1979	4,415	3,262	7,677
1980	4,518	3,273	7,791
1981	4,757	3,390	8,147
1982	4,534	3,379	7,913
1983	4,490	3,486	7,976
1984	4,825	3,663	8,488
1985	4,961	3,694	8,655
1986	4,994	3,803	8,797
1987	4,874	3,925	8,799

SOURCE: BPS.

labor between local/national varieties and HYVs around 1970. Collier and Birowo (1973) in a survey of 622 farmers from 20 Javanese villages found preharvest workdays per hectare for local and HYV varieties to be about the same (240). These findings were corroborated by the studies of Soelistyo (1975) in East Java and Montgomery and Sisler (1974) in the Yogyakarta area. Montgomery's surveys showed that both high-yielding and local varieties used 318 workdays per hectare. Soelistyo found that there was no significant difference between high-yielding and traditional varieties in labor use per hectare for preharvest activities in irrigated areas.

The shift from an open harvest with the *ani-ani* to a closed harvest with the sickle (yields were about the same among seed varieties in the early 1970s), however, resulted in a large drop in labor use per hectare with HYVs. Inferences from data in Collier's studies (1973, 1974) place the figure as high as 1,000 work hours per hectare in areas where farmers shifted from harvesting traditional varieties with an *ani-ani* to HYVs with a sickle.

As the decade of the 1970s passed and the yield and purchased-input gap between local and high-yielding varieties widened, labor requirements increased for some tasks of HYV cultivation. For example, fertilizer applications became more time-consuming as usage levels increased. Similarly, harvesting and threshing took longer as yields increased. But more generally, the time required to cultivate a hectare of wetland rice decreased. In a broad-ranging study of many villages, Collier et al. (1982a) estimated that labor-use requirements per hectare in wet rice cultivation including harvesting fell by 10 percent from 1969 to 1978. In a recent resurvey of villages and a reassessment of old evidence, Collier et al. (1988) estimated that labor use per hectare remained about the same through the 1970s but declined fairly substantially after 1980. The decline was attributed to decreased employment opportunities for women in harvesting, threshing, and weeding and for men in land preparation resulting from adoption of the technical innovations discussed above.

A number of problems arise from comparisons of such village-level studies of labor use patterns carried out at different times. Each study deals with peculiar ecological conditions and agronomic practices, size distribution of farms, enumeration techniques, and measurement errors. The regional coverage of the studies cited above varied markedly, and the average size of farms differed from one study to another. In all studies, data, especially on labor inputs in the traditional *ani-ani* harvest, are open to a considerable margin of error. These factors help to explain the huge range in estimates of labor inputs for each activity in the survey samples. Although useful and individually valid, the earlier survey work is not very helpful in quantifying the impact of the Green Revolution on total labor demand in the rice sector.

A different approach is taken here. Data on labor use per hectare from the field surveys are used to estimate current intensity in wetland rice agriculture on Java in 1987. Labor use per hectare in earlier periods is estimated by assessing the apparent impacts on labor requirements per task resulting from technical change, as documented in the work discussed above. Estimates of earlier labor use thus are obtained by working backward from recent survey evidence using old survey evidence as a guide. The aggregate impact on labor demand then is assessed by estimating the extent of the spread of various changes and multiplying by cultivated area. This scheme provides a rough estimate of the impact of the Green Revolution on labor use in Java—whether aggregate labor demand has increased, decreased, or remained about the same in Javanese rice agriculture.

Survey estimates of labor use per hectare by task per season for wetland rice in Java for 1987 are presented in Table 6.4. The table includes mechanized and nonmechanized cases and does not distinguish between hired and family labor. Total work hours per hectare through harvesting in 1987 range from 1,460, if all tasks are nonmechanized, to 1,010, if land preparation and threshing are mechanized.

As discussed above, land preparation in wetland rice cultivation on Java remains largely unmechanized. Estimates based on calculations from BPS survey data place tractor use at about 7 percent of total cultivated area in 1987 and 3 percent in 1980 (see above). Although data are not available, tractors are assumed not to have been used in 1969. Labor hours for both

TABLE 6.4. *Labor hours by task, wetland rice, Java (hours per hectare)*

Task	1969	1980	1987
Land preparation	300	300[a]	300[a]
		125[b]	125[b]
Planting	275	250	250
Weeding/water management	500	475	475
Fertilizing	0	20	30
Spraying	0	30	30
Harvesting	700[f]	350[c]	375[c]
		175[d]	200[d]
		275[e]	300[e]
TOTAL (nonmechanized)[a,c]	1,775	1,425	1,460

SOURCE: Author's estimates.
[a]Nonmechanized land preparation
[b]Mechanized land preparation
[c]Sickle harvest, manual threshing.
[d]Sickle harvest, machine threshing.
[e]Sickle harvest, pedal threshing.
[f]*Ani-ani* harvest.

mechanized and unmechanized land preparation (Table 6.4 assumes a combination of hand hoeing and animal traction for nonmechanized land preparation) in the earlier time periods are assumed to be the same as in 1987.

Planting was not subject to major changes over the time period. Minor technical innovations (e.g., spacing boards) appear to have lowered the time required to complete the planting task by 25 hours per hectare in 1980 and 1987 relative to 1969. Weeding done early in the cropping season when rice seedlings still are small was affected by the introduction of hand-pushed weeding tools. Time savings allowed by this innovation were offset somewhat by greater weed growth resulting from higher fertilizer use. Weeding time is assumed to have dropped by 25 hours per hectare in 1980 and 1987 in comparison with 1969. Because application of fertilizer and spraying of pesticides are associated with the spread of HYV seeds, labor requirements for these activities are set to zero in the 1969 estimation. Fertilizer application levels are higher in 1987 than in 1980, and that task is assumed to be 10 hours per hectare less time-consuming in 1980. According to the BPS cost of production surveys, pesticide usage levels did not change much between 1980 and 1987 so spraying time is assumed to be the same in both years.

The most important changes that have occurred in rice cultivation since 1969 are in harvesting. The shift from the *ani-ani* to the sickle allowed for considerable time savings in cutting paddy. These time savings have been offset very slightly by rising yields. The introduction of the sickle, *ceteris paribus*, has halved at least the number of labor hours required to harvest. Harvesting time with the sickle in 1980, then, is assumed to be half that in 1969 with the *ani-ani*. Because yields are higher, harvesting time per hectare is assumed to have risen 25 hours in 1987 in comparison with 1980. In 1987 harvesting required 375 hours per hectare. Harvest time is assumed to be 350 hours in 1980 and 700 hours in 1969. Adoption of the sickle on Java was estimated to have been carried out by at most 3 percent of wetland rice farmers in 1969, about 75 percent in 1980, and well over 90 percent in 1987. Data on adoption of the sickle apply to farmers rather than cultivated area and are considered to be rough estimates. For the estimation, sickles are assumed to have been used on 0 percent of cultivated area in 1969, 75 percent in 1980, and 100 percent in 1987.

Another technical change in harvesting is machine threshing. It is assumed that two-thirds of existing threshers are pedal-driven and one-third fuel-driven (the number of each type is not distinguished in the aggregate BPS data) and that pedal threshers are used for 10 hectares per season and fuel threshers 20 hectares per season. Each type of thresher is estimated (from agricultural survey data) to account for 2 percent of harvested area in 1980 and 8 percent of harvested area in 1987. Machine

TABLE 6.5. *Estimated annual labor demand, wetland rice agriculture, Java (000 labor hours)*

	1969	1980	1987
Annual labor demand	6,991,725	6,741,986	6,985,854
Percentage change from 1969		−1.6	−0.5

SOURCE: Author's calculations.

threshers were not present in Java in 1969. Labor hour estimates for machine threshing from the farm surveys are given in Table 6.4.

The results of the calculations of total labor demand are contained in Table 6.5. Annual labor demand in wetland rice cultivation on Java declined by 1.6 percent between 1969 and 1980. Labor demand in the 1980s has risen as a result of an expansion in cultivated area. Still, overall labor demand in 1987 is estimated to be 0.5 percent lower than it was in 1969. These results corroborate the conclusion from other work that on-farm labor demand in wetland rice production has not changed much on Java since the Green Revolution began. Increased demands, particularly from increased cropping intensities, have largely offset reduced demands for particular tasks (especially harvesting). Although stable in the aggregate, labor demands have become more evenly distributed through the year. The per capita employment obtained from rice probably has increased over the past two decades, whereas the total work force has declined.

Technical Change and Farm Income

The spread of the Green Revolution package of inputs and other related technical change (e.g., sickle harvesting) also affects the rural economy through changes in farm incomes. The full impact of changes in income will be discussed in Chapter 8. The task here is to estimate the changes in farm income from the production of rice that have occurred on Java as a result of the adoption of HYVs and infrastructural investments.

Both large- and small-scale farmers can use HYV seeds and chemical fertilizers profitably. Because of adoption of HYV seeds, increased fertilizer use, and better water control, rice yields have almost doubled in Indonesia's wetlands. In 1969 yields on *sawah* in Java were 2.6 tons of unmilled rice per hectare; by 1987, yields had increased to about 5 tons of *gabah* per hectare. Similarly, HYV seeds and irrigation investments enabled farmers to grow on average an additional half crop per year.

The labor coefficients reported in Table 6.4 are the starting point for the analysis. Data from the farm surveys are used to calculate on-farm profitability in 1987 for a representative one-hectare Javanese wetland rice farm using nonmechanized production technologies (i.e., tractors and machine threshers are not used). Nonlabor input use, wage, yield, and

TABLE 6.6. *Profitability of wetland rice cultivation on Java (nominal prices)*

	1969, traditional	1980		1987, HYV
		HYV	Traditional	
Yield (kilograms per hectare)	2,600	4,100	2,600	5,000
Output price (rupiahs per kilogram)	15	80	100	175
Gross revenue (000 rupiahs per hectare)	29	328	260	875
Labor costs[a]	11	115	112	352
Other input costs	1	28	7	68
Returns to land, management, and capital (000 rupiahs per hectare)				
Nominal prices	27	185	141	455
1969 prices	27	33	25	45
1987 prices	271	334	251	455

SOURCE: Author's estimates.
[a]Includes an imputed value to *bawon* harvest share.

output price data from various cost of production surveys published by the Central Bureau of Statistics are used to calculate on-farm profitability in 1969 and 1980.

This analysis indicates that annual incomes per hectare from rice cultivation in Java have almost tripled in real terms since 1969. These results are summarized in Tables 6.6 and 6.7. Table 6.6 shows the impact of higher yields only and indicates that rice profits (in constant 1987 prices) increased from Rp 271,000 per crop (per hectare) in 1969 for traditional seed varieties to Rp 334,000 per crop in 1980 and Rp 455,000 per crop in 1987 for high-yielding seed varieties. Table 6.7 includes the impact of higher average cropping intensities and indicates that on a per hectare basis annual real returns to wetland rice cultivation on Java have increased almost 150 percent since 1969. The rate of increase in income was greater in the 1980s than in the 1970s.

TABLE 6.7. *Wetland rice farming income, representative rice farm on Java*

	1969	1980	1987
Rice cropping intensity (rice crops per hectare)	1.25	1.62	1.84
Annual rice farming income (rupiahs per hectare)			
1969 prices	33,750	53,460	82,800
1987 prices	338,750	541,080	837,200

SOURCE: Author's estimates.

The Future of Labor Use in Rice Farming

During REPELITA V, if real wages continue to rise, mechanization of threshing and land preparation will spread more widely both on and off Java. Beyond profitability considerations, other influences could affect mechanization of rice agriculture in the future. One is the desire to reduce the drudgery involved in labor-intensive farming. Related to drudgery reduction is a desire on the part of farmers and laborers to raise the prestige of agricultural occupations. Both of these aspirations point in the direction of greater mechanization. Although the importance of these factors can only be guessed at, the 1987–88 survey found considerable sentiment in this regard, as have other researchers (Siregar 1986).

Changes that standardize cropping schedules in a particular region, thus raising the seasonal peaks in the demand for labor, also will affect the spread of future mechanization. Mechanization of land preparation has been induced in parts of West Java as a result of the synchronization of water-delivery schedules (Manning 1986). Similarly, tightness in inter-seasonal water schedules also induces mechanization; farmers must finish the previous crop quickly to be ready for the scheduled water delivery for the following one. In addition, the integrated pest management program will intensify seasonal peaks in labor demand through the simultaneous planting and harvesting of all contiguous *sawah* plots in an area.

The rate of mechanization will be influenced by farmers' ability to make investments in new technology. This is directly related to price and farmers' income. The cost of mechanical technology depends on government trade, tax, and credit policies. As discussed above, the present mix of government policies (e.g., trade taxes, supply monopolies, and credit policies) discriminates against the purchase of capital equipment. Government policies, therefore, do not encourage the mechanization of rice agriculture in Indonesia.

Conclusion

As shown in Chapter 2, farmers' returns are affected by government price policies for rice output and purchased inputs. Profits from rice production generally are very high (see Chapter 4). If high profits are maintained in the future, farmers, particularly those who cultivate good-control *sawah,* will continue to have high savings that could be used to purchase new farm implements. This ability to self-finance new capital investments will give them more flexibility to respond to rising wages.

Since the inception of the Green Revolution, there has been no change in aggregate employment in rice agriculture on Java. Real incomes in rice farming rose by two and one-half times per hectare between 1969 and

1987. Yet despite an increase in cultivated area, aggregate employment stayed about the same. Even if cultivated wetland rice area continues to increase in the future, input substitutions could cause rice production not to absorb many more farm laborers. Nevertheless, as shown in Chapter 5, rice farming done with mechanized technologies is still more labor-intensive than that of substitute crops.

7. Policy Alternatives for Future Rice Production Growth

Paul Heytens

Indonesia's efforts to increase rice production since the late 1960s have been enormously successful. Between 1969 and 1988, rice production grew at an annual rate of 4.5 percent. As a result, Indonesia moved from being the world's largest importer of rice in the 1970s to self-sufficiency in 1985. Successful development has been the result of policies that emphasized new technology, infrastructure investments, and remunerative prices to farmers. The dissemination of new and shorter-duration seed varieties, larger applications of fertilizer, better advice to farmers, and improved water control were key factors in increasing output. Rates of growth in rice output were roughly the same on and off Java and came mainly from increased productivity per hectare. Yields grew at about 3.3 percent annually, whereas harvested area increased only 1.2 percent annually.

Despite the past production success, continued future growth of rice output will require close attention by policymakers. Tabor et al. (1988) project that growth in rice consumption will remain above 2 percent per annum through the year 2000, although the rate of increase will decline from 2.6 percent a year in the early 1990s to 2.1 percent in the late 1990s. These consumption projections imply that rice production will need to grow at about 2.5 percent per annum through the remainder of REPELITA V for Indonesia to maintain self-sufficiency on trend. Although this rate is only half the growth in production attained during the past two decades, achieving a steady growth in rice production of 2.5 percent is no simple task. Much of the easier gains in both yields and irrigated area planted to rice already have been made.

Evidence from the field gives some reason for optimism that self-sufficiency can be sustained. Large-scale shifts into rice on *sawah* will not occur because rice already is being grown wherever feasible. But ample area could be transformed into higher-productivity systems through investments to improve water control. On Java, over 1 million hectares of rainfed and poor-control, irrigated *sawah* could be transformed by investments in new irrigation systems or rehabilitation of existing systems. Field evidence and the historical record in Indonesia indicate that beneficiaries of better-controlled water supplies respond with increased input use to exploit new opportunities. Further, to the extent that agricultural research and extension services can make available new technologies such as integrated pest management and hybrid rice seeds Indonesia's rice farmers can be expected to adopt innovations that will allow them to attain higher profitability and productivity. Finally, Java's sugar producers would move out of cane into rice-dominated cropping patterns if the government allowed them to do so.

The desirability of making investments in irrigation infrastructure, maintaining incentives for farmers and a vital research and extension program, and deregulating sugar production depends partly on how efficiently Indonesia is able to produce rice and competing crops. Do increases in rice output, needed for greater food security, also contribute to efficient growth of national income, or is there a trade-off between food security and efficiency? Does Indonesia have a comparative advantage in producing rice for import substitution or for export? Are there competing crops that provide higher social profits than those for rice? The answers to these questions, addressed below, go a long way in defining an appropriate rice strategy for Indonesia.

Social Profitability of Rice and Competing Crops

To assess Indonesia's comparative advantage, the social profitability of rice and competing crops must be calculated. For that purpose, policy analysis matrices (PAMs) were constructed. This method of analysis simultaneously addresses the prospective changes in levels of rice income, the contribution of rice to efficient growth of national income, and the income transfers resulting from price and macroeconomic policies. The main task is to construct, for each rice production system defined in Chapter 4, a set of accounting matrices of revenues, costs, and profits, first including and then excluding the effects of policy and market failures. The approach is described in detail in Monke and Pearson (1989).

The first part of this task was accomplished by the construction of budgets from data gathered during the farmer surveys. The second part requires revaluing private costs and benefits at social or efficiency values—what costs and returns would be if there were no distorting policies or market failures. The world price of any tradable input or output, with appropriate adjustments for location and quality, represents the price at which inputs and outputs could be bought or sold without intervening policies. For example, in the absence of intervening policy, farmers could purchase an input traded on world markets for the world price (c.i.f. at the nearest port) plus the cost of transport to the farm gate. Similarly, traded output (which includes rice) could be sold for the world price (c.i.f. at a domestic port for an import substitute or f.o.b. for an export). Assessing the social values for inputs that are not traded (e.g., transport, capital, and labor) is more difficult, but the principle remains the same as with tradables—to discern costs in the absence of distorting policies and market failures.

For this analysis, the production budgets presented in Chapter 4 are generalized to represent wetland systems for Java, South Sulawesi, and West Sumatra, the project's survey regions. Private budgets estimated from survey results were checked for consistency against data from the triannual cost of production surveys done by Indonesia's Central Bureau of Statistics (BPS)–that is, a weighted average of the rice production budgets described in Chapter 4 was estimated to verify its similarity to the BPS average budget for Indonesia.

Policy Analysis Matrix Assumptions

The PAM approach requires evaluation of costs and benefits at social or efficiency values. In practice, this task is based on a revaluation of private costs and benefits and involves making judgments about key parameters. Therefore, if the PAM analysis is to be a useful guide to Indonesia's policymakers, assumptions for important parameters must be made explicit and sensitivity analysis done to assess the robustness of results to different assumptions.

For this analysis, social values for rice output, the foreign exchange rate, the interest rate, and labor costs are particularly important. Domestic rice prices were in line with world prices of comparable quality grains in 1987 (c.i.f. Indonesian ports of about $200 per metric ton for Thai 35 percent brokens). Although prices on the world rice market are very difficult to project (as Chapter 3 makes clear), world rice prices in 1987 were judged to be below their expected long-run trend. A c.i.f. Indonesia price of $250 per metric ton consequently is used as the social value of Indonesian rice in the base case.

As described in Chapter 2, the history of Indonesian exchange rate policy in the 1970s and 1980s is characterized by periods of overvaluation caused by positive differential inflation (i.e., domestic inflation higher than that of major trading partners) followed by large corrective devaluations. Although differential inflation was positive in 1987 and import restrictions created some unrealized demand for foreign exchange, any overvaluation of the rupiah then was not large. A major devaluation of the rupiah in September 1986 probably brought the Indonesian currency back into line with its opportunity cost at that time, and any subsequent overvaluation during 1987 was offset by the sharply rising value of the yen relative to the dollar (and thus the rupiah). Therefore, as a base assumption, no correction is made to the private Rp/U.S.$ exchange rate.

The private return to capital does require some adjustment for social valuation of capital costs. Credit policy differs from that in other developing countries because the private real interest rate (14 percent) is higher than its social counterpart. This supposition is corroborated by the 4 to 5 percent decline in nominal interest rates that occurred after the implementation of the financial sector deregulation package in late 1988. Other analysis indicates that a real social rate of return of 10 percent is a reasonable approximation for a country such as Indonesia (Monke and Pearson 1989).

In the calculation of social costs, the 10 percent rate is applied to purchases of large capital equipment such as tractors and large trucks for transport. It is not intended as the social opportunity cost of farmers' working capital. Evidence from the KUPEDES (rural credit) program indicates that the interest rate at the village level for small savers and borrowers is considerably higher than the rate in urban centers. Nominal KUPEDES savings rates were in the 30 percent range in 1987. Given an inflation rate of 8 to 9 percent, the real social opportunity cost of farmers' cash outlays is considered to be 20 percent per year in the base case.

Finally, a judgment must be made about the competitiveness of Indonesia's rural labor markets. Chapter 5 provides evidence that the labor market is operating with reasonable efficiency. Therefore, no divergence is assumed between private and social labor costs in the analysis. A complete listing of parameter values used and other assumptions made is found in the appendix at the end of the chapter.

Policy Analysis Matrix Results

The results of the social and private profitability calculations for rice and competing crops for the base year 1987 are reported in Table 7.1. The table shows that rice as an import-substitution activity (i.e., measured relative to a c.i.f. Indonesia world price) is socially profitable on all types of *sawah* in the three regions. Further, social profits for rice are higher than for competing *palawija* crops, and the social returns to land and management for soybeans and sugarcane are negative. A move toward world prices or a reduction in planted area targets for either of these commodities, but particularly for sugar, would free up additional cropland for the production of more socially profitable rice. This efficient change would increase farm incomes, food production, and national income.

Government policy effects a relatively small magnitude of income transfers to rice producers through tradable inputs, as indicated in Table 7.2. The source of policy transfers for tradable inputs is the direct subsidy on chemical fertilizers and pesticides, and transfers are largest in the most productive systems. But the magnitudes involved are small. The fertilizer subsidy varies by type but is largest for urea and triple super phosphate, the most heavily used tradable inputs.

TABLE 7.1. *Policy analysis matrix, wetland rice and competing crops (000 rupiahs per hectare)*

System	Private				Social			
	Revenues	Tradable inputs	Domestic factors	Returns to land and management	Revenues[a]	Tradable inputs	Domestic factors	Returns to land and management
Java								
Good-control, wet	1,341	110	491	740	1,670	156	545	969
Good-control, dry	1,523	119	536	868	1,670	176	584	910
Moderate-control, wet	1,126	108	458	560	1,403	161	515	727
Moderate-control, dry	1,219	101	489	629	1,336	148	534	654
Poor-control, wet	912	91	476	345	1,135	133	509	493
Poor-control, dry	1,036	86	515	435	1,135	126	534	475
Rainfed	858	85	458	315	1,069	125	467	477
South Sulawesi								
Moderate-control, wet	1,126	132	380	614	1,403	192	356	855
Moderate-control, dry	1,158	129	384	645	1,269	189	360	720
Poor-control, wet	912	78	350	484	1,135	114	343	678
Poor-control, dry	853	74	349	430	935	110	343	482
Rainfed	751	69	319	363	935	96	303	536
West Sumatra								
Good-control, wet	1,341	131	715	495	1,670	176	704	790
Good-control, dry	1,523	131	728	664	1,670	176	714	780
Moderate-control, wet	1,019	116	659	244	1,269	161	659	449
Moderate-control, dry	1,158	116	669	373	1,269	161	668	440
Poor-control, wet	804	69	651	84	1,002	101	665	236
Rainfed	697	51	584	62	868	61	584	223
Java								
Soybeans, irrigated	683	89	326	268	416	117	347	−48
Yellow corn, irrigated	700	109	278	313	749	160	296	293
Sugar, irrigated[b]	4,800	345	2,857	1,598	2,632	421	2,753	−542

SOURCE: Author's calculations.
[a]All revenues are based on c.i.f. import prices; see Appendix 7.1 for price information.
[b]Farmers receive only 62 percent of gross revenues but pay over 80 percent of input costs; the remainder accrues to processors.

TABLE 7.2. *Policy transfers and market failures, 1987 (000 rupiahs per hectare)*

System	Output	Tradable inputs	Domestic factors	Total
Java				
Good-control, wet	−329	46	54	−229
Good-control, dry	−147	57	48	−42
Moderate-control, wet	−277	53	57	−167
Moderate-control, dry	−117	47	45	−25
Poor-control, wet	−223	42	33	−148
Poor-control, dry	−99	40	19	−40
Rainfed	−211	40	9	−162
South Sulawesi				
Moderate-control, wet	−277	60	−24	−241
Moderate-control, dry	−111	60	−24	−75
Poor-control, wet	−223	36	−7	−194
Poor-control, dry	−82	36	−6	−52
Rainfed	−184	27	−16	−173
West Sumatra				
Good-control, wet	−329	45	−11	−295
Good-control, dry	−147	45	−14	−116
Moderate-control, wet	−250	45	0	−205
Moderate-control, dry	−111	45	−1	−67
Poor-control, wet	−198	32	14	−152
Rainfed	−171	10	0	−161
Java				
Soybeans, irrigated	267	28	21	316
Yellow corn, irrigated	−49	51	18	20
Sugar, irrigated	2,168	76	−104	2,140

SOURCE: Author's calculations.
NOTE: Positive numbers indicate subsidies and negative numbers denote taxes.

The major policy transfers in the domestic factors category result from subsidies on transport and handling of fertilizers and pesticides to the village cooperative (KUD) and the free provision (generally) of irrigation water. The transport subsidy for fertilizer and pesticides is of roughly the same magnitude in all regions, but the irrigation subsidy is considerably larger on Java than in the outer islands. In addition, divergences in urban capital markets have the effect of increasing private tractor rental and transport costs for rice output above social costs. The magnitude of factor market transfers, however, is small and does not offset the overall positive effects on farmers' incomes of the tradable input subsidies.

On the output side, Table 7.2 indicates that government policy is taxing rice producers. As discussed above, world rice prices in 1987 were judged to be below long-run trend values. Since domestic rice prices in 1987 were broadly in line with actual world prices, there is an apparent tax on producers. This implicit tax arises because government policy forced rice producers to accept a domestic price that was below the trend world rice price in 1987. The difference is about $50 per metric ton plus domestic transport costs, more than 25 percent of the social value of rice.

Sensitivity Analysis

Rice production's comparative advantage as an import-substitution activity is very robust to changes in parameter assumptions. For example, adjustments in the foreign exchange rate to reflect a possible overvaluation only strengthen Indonesia's comparative advantage because a higher Rp/U.S.$ exchange rate serves to raise the rupiah cost of competing rice imports and hence the rupiah value of domestic rice. In addition, doubling the real social rate of return to 20 percent and increasing the opportunity cost of farmers' working capital to 30 percent raises domestic factor costs by not more than Rp 50,000 per hectare for any of the rice systems studied, hardly enough to challenge base case conclusions.

The more interesting parameters for sensitivity analysis are the break-even prices for import and export competitiveness, that is, the minimum c.i.f. and f.o.b. prices that would have to prevail for Indonesian rice to be competitive. Under the base case assumptions, the c.i.f. price for Thai 35 percent brokens could fall to roughly $180 per metric ton before Indonesia's least efficient wet-rice systems would cease to be socially profitable. Furthermore, Indonesia might be able to export rice profitably when world prices are at or above long-run trend values (f.o.b. prices at or above $200 per metric ton). Because of problems associated with the thin world rice trade, the depressing impact of Indonesian sales on export prices, the quality of Indonesian rice, and the long gestation effects of investments related to market development, exports are more problematic and the results must be interpreted more cautiously. But the analysis indicates at least that occasional exports in times of domestic surplus could be done efficiently. The future choice of domestic price levels needs to consider the possibility of efficient exports because price support levels, not production costs relative to f.o.b. prices, will determine whether Indonesian exports can compete without subsidy.

Defining a Future Production Strategy

In 1987, Indonesia had a strong comparative advantage in producing rice as an import substitute. At that time, it appeared economically efficient to invest resources in rice agriculture to maintain trend self-sufficiency. Not doing so, and importing rice, would have been an inefficient use of resources. This result is strong evidence against the adoption of a near-term future rice production strategy that would imply a low rate of growth of output (about 1 percent annually) and lead to regular imports. In the remainder of this chapter, therefore, the low production target is ignored, and attention is focused instead on the two other output targets and their associated strategies—2.5 percent annually to achieve self-sufficiency on trend and 4 percent annually to support absolute self-sufficiency and regular exports. The central issue is whether achievement of either of these two growth targets would be consistent with the objectives of food security and efficient income generation. Their likely impacts on equity (personal and regional income distribution) are examined in Chapter 8.

Public Investment in Irrigation

REPELITA V provides estimates of the public-sector investments that are likely to take place by early 1994.[1] Of particular importance for this study are the irrigation investments that will convert unused or low-productivity land into higher-yielding systems. Transformation of rainfed land involves the building of new irrigation works, whereas upgrading poor- and moderate-control *sawah* requires rehabilitation of already existing infrastructure. In addition, expenditures on operations and maintenance of the entire system are needed to prevent physical deterioration of past investments and the consequent loss of rice output.

The Fifth Plan allocates Rp 8.5 trillion to the irrigation subsector between 1989 and early 1994. The physical targets include about 500,000 hectares of new irrigation development, 450,000 hectares of upgraded swamp irrigation, and 450,000 hectares of additional flood protection. REPELITA V also calls for various operation and maintenance expenditures on nearly 6 million hectares.

Irrigation projections for REPELITA V are based on the actual experiences of REPELITAS I through IV. Table 7.3 outlines financial and physical data for four key irrigation programs between 1969 and 1988. The data on plans versus actual realization show considerable variation by category and time period; however, these performances and experiences are consistent with those encountered by other countries in Asia. Without considerably more fieldwork to verify specific local investments, even the most carefully made projections are likely to provide only rough approximations of irrigation's contribution to rice growth during the next five years.

One such approximation is given in Table 7.4. Each of the investment categories is assessed relative to its REPELITA V target. Each category then is broken down into intensity, area, and yield components. The assumptions for each type of investment are made explicit in Table 7.4 so that analysts who have alternative views on irrigation investments during REPELITA V and the lagged effects that could accrue from investments completed during earlier plans can produce their own assessments.

[1] This section draws heavily from Varley (1989).

TABLE 7.3. *Plans and realization, irrigation subsector, REPELITAs I through IV (000 hectares and billion rupiahs)*

	REPELITA I		REPELITA II		REPELITA III		REPELITA IV		Average achievement (percent)
	Plan	Realized	Plan	Realized	Plan	Realized	Plan	Realized	
Rehabilitation									
Expenditure	50	50	176	148	673	556	1,328	754	71
Area	830	836	585	528	536	394	360	200	85
New irrigation investment									
Expenditure	25	25	456	197	752	760	3,361	1,549	55
Area	430	191	950	326	700	436	600	218	44
River/flood control									
Expenditure		6	203	219	680	482	1,863	951	60
Area		289	679	614	770	578	500	364	95
Swamps									
Expenditure		33			208	110	278	74	45
Area		119			535	476	494	181	75

SOURCE: Data from Varley (1989).

TABLE 7.4. *Impact during REPELITA V of irrigation investments in* gabah *production*

	Target area in REPELITA V (million hectares)	Estimated achievement (million hectares)	Increased crop intensity	Yield increment per unit area (metric tons per hectare)	New harvested (million hectares)	Increase from new area (million metric tons)	Increase on existing area (million metric tons)	Total incremental production (million metric tons)
Upgrading of swamp	0.444	0.200	0.10	1.00	0.020	0.046	0.200	0.25
Conventional rehabilitation								
Java	0.222	0.150	0.10	0.50	0.015	0.081	0.116	0.20
Off Java	0.112	0.075	0.10	0.50	0.008	0.030	0.053	0.08
Intensification	0.500	0.300						
Existing projects	0.100	0.100	1.50	4.00	0.150	0.600	0.000	0.60
New land development	0.200	0.100	1.25	4.00	0.125	0.500	0.000	0.50
Tertiary/secondary	0.200							
Java		0.025	1.65	5.00	0.041	0.206		0.21
Off Java		0.075	1.50	4.00	0.113	0.450		0.45
Loss of output on rainfed land								
Java		0.025	1.00	3.50	0.025	0.088		(0.09)
Off Java		0.075	1.00	3.00	0.075	0.225		(0.22)
Incremental increased *gabah* production from irrigation program								1.97

SOURCE: Data from Varley (1989).

Four important conclusions follow from the assumptions and results of this projection. First, continued investments in irrigation should produce substantial further growth in rice output. Specifically, a 2 million ton increase in *gabah* production seems plausible by the end of 1993 from irrigation improvements. This tonnage translates into about a 1 percent per year increase in rice output during REPELITA V. Second, the projections in Table 7.4 imply no production increases from operation and maintenance expenditures on irrigation systems. Keeping irrigation systems at reasonable efficiency is a continuing problem. For irrigation, as well as for new seeds, Indonesia will require major expenditures and organizational efforts simply to maintain previous gains in rice output. Third, the gains from new irrigation investments will not be captured so easily as those from earlier investments. Many of the easiest systems have been improved in earlier plans, leaving the more remote and costly irrigation projects for the 1990s. Finally, the resurgence in actual irrigation investments since 1986 and the proposed investments in REPELITA V are good indicators of the importance the government attaches to rice self-sufficiency and of the key roles played by irrigation in such a strategy.

Regulatory Policy

Perhaps the quickest way to increase rice production during REPELITA V would be to change sugar policy. Roughly 150,000 hectares of *sawah* were in sugarcane on Java in 1988, mostly in the TRI program (Nelson 1988). If the government were to phase out the TRI program, perhaps over a five-year period, and current relative prices were to hold, about 150,000 hectares of *sawah*, mostly in moderate- and good-control irrigated systems in East and Central Java, could shift into rice production.

An informal phaseout of TRI sugarcane land began in Java in 1989, because program regulations were not enforced in some parts of Java. For purposes of estimation, a phaseout plan is modeled so that 30,000 additional hectares of sugarcane land are planted with other crops each year for five years beginning in June 1989. In 1989, two dry-season crops per plot could be grown on the initial 30,000 hectares. In the first dry season, rice is assumed to be grown on four-fifths and *palawija* on one-fifth of the newly converted *sawah*. In the second dry season, two-fifths of the available land is assumed to go into rice, two-fifths into *palawija* crops, and one-fifth into fallow rainfed *sawah*. In 1989, then, an additional 36,000 hectares of rice (and 18,000 hectares of *palawija*) would come into production. In subsequent years, it is assumed that 40 percent of the converted sugar land could accommodate three rice crops, 40 percent two rice crops, and 20 percent only one rice crop. In 1990, for example, cultivated rice area would increase by 102,000 hectares. Land changed over in 1989 would provide 66,000 hectares of rice production (plus 18,000 hectares of *palawija* production), and 36,000 hectares would be converted anew to rice. The implications of such a program for rice production are summarized in Table 7.5.

The table indicates that the sugarcane phaseout under the above assumptions would raise cultivated rice area on *sawah* by 330,000 hectares by 1994 because of multiple cropping. If yields on converted land average 5.25 tons of *gabah* per hectare in 1989 and increase by 0.5 percent annually, the phaseout would result in about 1.8 million tons of additional *gabah* production per year by 1994. In comparison with production of over 41 million tons of unmilled rice in 1988, the production changes shown in Table 7.5 imply, *ceteris paribus*, annual growth in rice production of 1 percent through the end of REPELITA V.

Price Policy

Price policy is the residual lever that Indonesia can use to affect rice production during REPELITA V. If a production target needed to realize a chosen strategy cannot be achieved with changes in investment and regulatory policy, price policy is the last option available to provide additional incentives to rice farmers. Altemeier et al. (1988) estimate that the price elasticity of rice yields is 0.2 in the first year and 0.3 after three years if the price change persists. Hence if the real rice price increases 10 percent and is held at that level, rice yields can be expected to increase about 2 percent after one year and a total of 3 percent over a three-year period. Price policy influences cultivated area as well. Altemeier et al. (1988) estimate the one-period lagged own price elasticity for rice area to be 0.17.

The rapid increases in real prices during 1987 and 1988 have encouraged area expansion and intensification of input usage in rice, which led to a strong rebound in production early in REPELITA V. The roughly 20 percent real price rise from 1986 to 1988 should generate yield growth of

TABLE 7.5. *Projected rice production, sugarcane phaseout, 1989–94*

Year	Cumulative cultivated rice-area added (000 hectares)	Projected yield (tons per hectare)	Additional annual *gabah* production (tons)
1989	36	5.25	189,000
1990	102	5.28	538,177
1991	168	5.30	890,842
1992	234	5.33	1,247,020
1993	300	5.36	1,606,737
1994	330	5.38	1,776,248

SOURCE: Author's calculations.

FIGURE 7.1. *Movements in the price of "medium" rice, Indonesia, 27 cities, weighted average* (from BULOG)

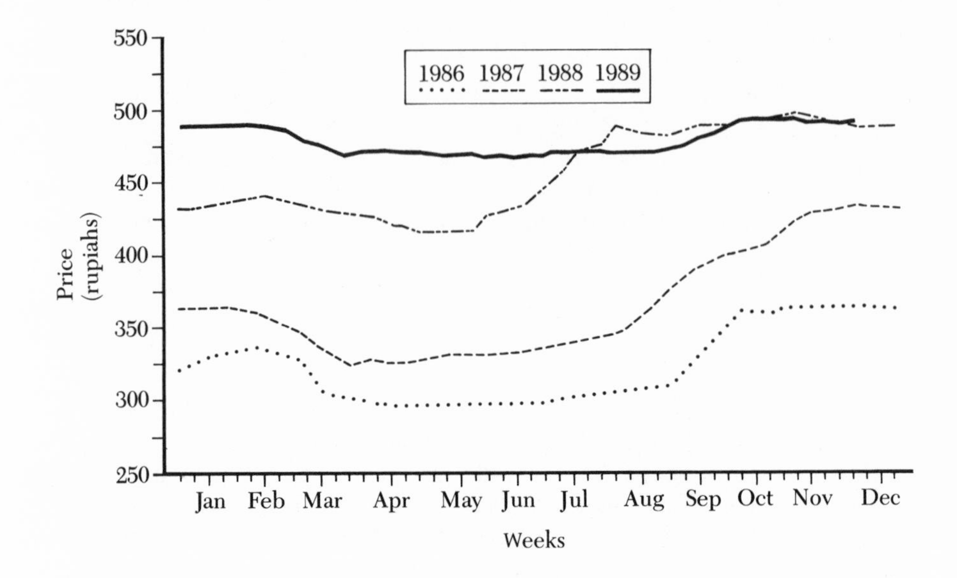

about 1 percent in 1989 and about 0.5 percent in 1990. Harvested rice area probably increased by nearly 2 percent in 1989 in response to the earlier price increases. If real prices were held constant throughout the REPELITA V period, the effects of the earlier price increase would stop contributing to rice production growth by the end of 1990.

Figure 7.1 shows the movements of nominal rice prices, by month, between 1986 and 1989. The rapid increase in prices occurred between August 1987 and August 1988, reflecting drought and a decision not to import rice. This abnormally large increase moved Indonesian domestic prices from only 80 percent of expected long-run world price levels to approximate equality.

Between August 1988 and late 1989, nominal domestic prices moved very little, and real prices fell somewhat (the annual rate of inflation was about 6 percent during this period). The stability in prices resulted from excellent weather, repayments of rice "loans" from the Philippines and Vietnam, changes in the enforcement of the TRI sugar policy (discussed above), and price responsiveness of rice farmers. This combination of good fortune and policy permitted BULOG to rebuild its food reserve stock and the country to regain aggregate supply-demand balance in rice. By the close of 1989, then, the Indonesian rice economy had attained trend self-sufficiency. But policymakers had very little unused efficient policy space within which to raise real domestic prices.

Determining a Policy Mix for REPELITA V

Policymakers are concerned with the changes in rice policies required to pursue a rice strategy, such as trend self-sufficiency or regular exports. The first step in this assessment is to quantify the increases in rice production during the Fifth Plan period that are already in the pipeline. The sources of growth include carryover effects from price changes in earlier years, the impact of public investments (both lagged and future) in irrigation infrastructure, the impact of reductions in sugarcane area already made on Java, and trend effects attributable to improved practices apart from the irrigation program. The potential contribution of these factors to rice production during REPELITA V is summarized in the first column of Table 7.6.

Table 7.6 incorporates the carryover effects from the rice price increases of 1986–88 as well as the phase-in of the incremental *gabah* production resulting from irrigation investments, shown in Table 7.4. In addition, the first year of the sugarcane phaseout plan described above is assumed to have been implemented informally. Moreover, a 0.5 percent per annum growth in yields, additional to the impact of irrigation investments, is assumed to incorporate the effects of wider implementation of integrated pest management, the spread of other intensification programs, and local innovations. No trend effect outside of irrigation investment is assumed for rice area. The trend assumptions are low in comparison with past achievements, reflecting the existing high rates of HYV adoption and intensive use of modern inputs, especially chemical fertilizers.

The results in the first column of Table 7.6 show that, without additional policy incentives, a sharp downturn in production growth would occur after 1989. Annual growth during the final four years of REPELITA V averages somewhat over 1 percent per year. To avoid consistent rice imports during REPELITA V, then, Indonesia likely would have to implement additional incentive policies. The policy analysis matrix results (Table 7.1) indicate that Indonesia has a strong comparative disadvantage in sugarcane production. Irrespective of the rice strategy Indonesia ultimately chooses, it is clearly in the country's economic best interest to continue with the sugarcane phaseout summarized in Table 7.5. The implications of following through with the phaseout of the TRI sugarcane policy are summarized in the middle column of Table 7.6. That column shows that growth in rice production increases to over 2 percent per year

TABLE 7.6. *Projected rice production, 1988–93*

	Base case				Base case + continued sugarcane phaseout				Base case + sugarcane phaseout + 5 percent real price increase (1991)			
	Yield (tons per hectare)	Area (hectares)	*Gabah* production (tons)	Annual growth (percent)	Yield (tons per hectare)	Area (hectares)	*Gabah* production (tons)	Annual growth (percent)	Yield (tons per hectare)	Area (hectares)	*Gabah* production (tons)	Annual growth (percent)
1988												
TOTAL	4.11	10,138,155	41,667,817		4.11	10,138,155	41,667,817		4.11	10,138,155	41,667,817	
1989												
Trend and price	4.17	10,313,545	43,027,685		4.17	10,313,545	43,027,685		4.17	10,313,545	43,027,685	
Sugar	5.25	36,000	189,000		5.25	36,000	189,000		5.25	36,000	189,000	
Irrigation			400,000				400,000				400,000	
TOTAL			43,616,685	4.68			43,616,685	4.68			43,616,685	4.68
1990												
Trend and price	4.21	10,313,545	43,459,037		4.21	10,313,545	43,459,037		4.21	10,313,545	43,459,037	
Sugar	5.28	66,000	348,480		5.28	102,000	538,177		5.28	102,000	538,177	
Irrigation			800,000				800,000				800,000	
TOTAL			44,607,517	2.27			44,797,215	2.71			44,797,215	2.71
1991												
Trend and price	4.23	10,313,545	43,676,333		4.23	10,313,545	43,676,333		4.28	10,313,545	44,113,096	
Sugar					5.30	168,000	890,842		5.30	168,00	890,842	
Irrigation			1,200,000				1,200,000				1,200,000	
TOTAL			44,876,333	0.60			45,767,175	2.17			46,203,938	3.14
1992												
Trend and price	4.26	10,313,545	43,894,714		4.26	10,313,545	43,894,714		4.31	10,402,757	44,828,940	
Sugar					5.33	234,000	1,247,020		5.33	234,000	1,247,020	
Irrigation			1,600,000				1,600,000				1,600,000	
TOTAL			45,494,714	1.38			46,741,734	2.13			47,675,960	3.19
1993												
Trend and price	4.28	10,313,545	44,114,188		4.28	10,313,545	44,114,188		4.34	10,402,757	45,165,718	
Sugar					5.36	300,000	1,606,737		5.36	300,000	1,606,737	
Irrigation			1,970,000				1,970,000				1,970,000	
TOTAL			46,084,188	1.30			47,690,925	2.03			48,742,455	2.24

SOURCE: Author's calculations.

during the final three years of the Fifth Plan, but not enough to reach 2.5 percent, the likely rate needed to attain self-sufficiency on trend.

Some increase in incentives to farmers seems necessary as well, although budgetary pressures make high fertilizer subsidy levels difficult to maintain. As a result, administered fertilizer prices were raised considerably for 1990 (see Table 2.1). The recent series of fertilizer price increases should not have a large disincentive impact on rice production because Indonesia's wetland rice farmers have had long experience using chemical fertilizers and application rates are already very high (Altemeier et al 1988). Further, there is strong agronomic evidence in Indonesia that TSP and KCl applications could be reduced substantially with virtually no impact on rice yields (Manwan and Fagi 1989; Adiningsih et al 1989) but with considerable savings for the government budget.

If output price policy is used, a 5 percent real price increase in 1990 or 1991 would be required to achieve rice self-sufficiency on trend during REPELITA V. The impact of a 5 percent price increase in 1991 is summarized in the final column of Table 7.6. The projected production response would place Indonesia somewhat above 2.5 percent annual growth through REPELITA V and would allow policymakers the possibility to continue phasing out fertilizer subsidies.

Higher annual growth rates in rice production than these, of course, would require even higher real rice prices (in the assumed absence of even greater investment or faster technical change). For example, an average annual growth rate of 4 percent after 1989 would require an additional real increase in rice prices of at least 10 percent during REPELITA V. Because real rice prices rose sharply during the 1986-88 period, potential political difficulties are likely to impede further large increases in the near term. Higher domestic rice prices also make Indonesian exports less competitive. To generate exportable surpluses, policymakers would have to raise domestic rice prices considerably, but this high price policy would cause the exported quantities to require subsidies in most years. Therefore, pursuit of the high-growth rice strategy probably is not desirable, unless it could be achieved with public investments and technical changes additional to those considered here.

A final caveat is in order. This chapter has focused on trends and longer-run orders of magnitude. Short-term variations in international rice prices, domestic rice production, or Indonesian oil revenues could create opportunities or constraints for rice policy that planners cannot foresee in advance. A successful rice strategy for the 1990s needs to be based on sound long-run planning and investments. It also requires short-run flexibility and the capacity to adjust policy to changing year-to-year circumstances.

Conclusion

In comparison with either the actual 1987 world rice prices or the expected, long-run trend prices, Indonesia produces rice efficiently to substitute for imports. Comparative advantage in import substitution is strongest in the good-control, irrigated systems, but exists in all wetland rice production (about 90 percent of total production). Because real domestic rice prices rose 20 percent between mid-1987 and mid-1988, however, policymakers cannot afford to allow domestic levels to rise much more if they wish to maintain efficiency. World rice prices are expected on average to fluctuate around a long-run trend price of about $250 per metric ton (c.i.f. Indonesian ports, for 35 percent brokens). At the end of 1989, Indonesian prices were approximately aligned with this expected trend of c.i.f. import prices. Consequently, policymakers cannot use large price policy incentives to induce efficient increases in rice output and farmers' incomes; production achieved at domestic prices in excess of comparable world prices is inefficient and wastes scarce resources. In this circumstance, Indonesia faces an emerging trade-off between the objectives of improved food security and the efficient generation of income.

The analysis of likely sources of growth for rice production shows that this trade-off could emerge as soon as 1991. By then, the residual incentive effects of the price rise in 1987–88 will have disappeared. Moreover, the projected increases in rice output from public investments in irrigation, research, and other infrastructure and from phasing out the TRI sugarcane program over five years amount to just over 2 percent annually. By 1991, therefore, price policy incentives could be needed to permit yearly growth in rice production of 2.5 percent, the projected rate of increase of rice consumption.

The implications of the analysis are clear. A strategy of regular imports (based on a 1 percent annual growth target) is inappropriate; Indonesia can compete efficiently with rice imports at expected c.i.f. prices. At the other extreme, a strategy of planning for regular rice exports with a high production target (about 4 percent annually) is also likely to be ill-advised. Given the recent (and expected long-run) relationships between Indonesian domestic prices and export prices, the country cannot sell regularly on world markets without export subsidies; this lack of export competitiveness would be exacerbated if Indonesia chose to raise real domestic rice prices substantially in an effort to generate exportable supplies. The most desirable strategy, from the viewpoints of both food security and efficiency, is self-sufficiency on trend. But even this strategy faces potential efficiency problems if substantially higher domestic prices are required to balance domestic supply and demand.

Appendix 7.1. Policy Analysis Matrix Assumptions

Following are the assumptions used for the policy analysis matrix analysis as well as tables of private and social values for general and regional prices and system input use and output data. Important assumptions are justified in the text.

General parameter values used

	Private	Social
Exchange rate	Rp 1,644 per U.S.$	Rp 1,644 per U.S.$
Opportunity cost of capital equipment	14% per year	10% per year
Opportunity cost of working capital	30% per year	20% per year

Output prices used (rupiahs per kilogram)

	Domestic	World
Wet-season rice	330	411
Dry-season rice	375	411
Yellow corn	200	214
Soybeans	700	429
Sugar	600	329

Java rice systems, labor and input use

	Good-control *sawah*		Moderate-control *sawah*		Poor-control *sawah*	
	Wet season	Dry season	Wet season	Dry season	Wet season	Dry season
Labor task (hours per hectare)						
Manual land preparation	200	200	200	200	250	250
Animal plowing	70	70	70	70	100	100
Planting	250	250	250	250	250	250
Weeding and water management	450	575	450	575	525	650
Fertilizing/spraying	60	56	66	50	75	60
Harvesting	425	425	400	400	350	350
Drying	24	16	24	16	24	16
Fertilizer (kilograms per hectare)						
Urea	200	250	250	250	250	250
TSP	100	150	100	100	75	75
KCl	50	50	25	25		
ZA	50	50	25	25		
Other inputs						
Liquid insecticide (liters per hectare)	1.5	1	2	1	2.5	1.5
Granulated insecticide (kilograms per hectare)	8	6	12	8	6	4
Seed (kilograms per hectare)	35	35	35	35	35	35
Spraying (times per hectare)	4	3	5	4	5	4
Yield (kilograms per hectare)	6,250	6,250	5,250	5,000	4,250	4,250

South Sulawesi rice systems, labor and input use

	Moderate-control *sawah*		Poor-control *sawah*		
	Wet season	Dry season	Wet season	Dry season	Rainfed *sawah*
Labor task (hours per hectare)					
Tractor rental	18	18	–	–	–
Animal plowing	–	–	45	45	60
Planting	100	100	100	100	100
Weeding and water management	85	135	100	180	110
Fertilizing/spraying	21	21	41	41	15
Harvesting	300	280	270	260	260
Drying	24	8	24	8	16
Fertilizer (kilograms per hectare)					
Urea	200	200	200	200	200
TSP	100	100	50	50	–
KCl	50	50	–	–	–
Other inputs					
Liquid insecticide (liters per hectare)	–	–	2	2	–
Granulated insecticide (kilograms per hectare)	20	20	–	–	10
Herbicide (liters per hectare)	2	2	1	1	1
Seeds (kilograms per hectare)	30	30	30	30	30
Spraying (times per hectare)	–	–	4	4	–
Yield (kilograms per hectare)	5,250	4,750	4,250	3,500	3,500

Java *palawija* systems, labor and input use

	Irrigated soybeans	Irrigated yellow corn
Labor task (hours per hectare)		
Manual land preparation	60	60
Animal plowing	80	80
Planting	200	60
Weeding and water management	420	140
Fertilizing/spraying	54	90
Harvesting	210	300
On-farm processing	155	24
Other inputs		
Urea (kilograms per hectare)	200	450
TSP (kilograms per hectare)	–	50
Liquid insecticide (liters per hectare)	4	–
Granulated insecticide (kilograms per hectare)	–	12
Spraying (times per hectare)	4	–
Seed (kilograms per hectare)	50	35
Yield (kilograms per hectare)	975	3,500

West Sumatra rice systems, labor and input use

	Good-control *sawah*, wet and dry	Moderate-control *sawah*, wet and dry	Poor-control *sawah*, wet season	Rainfed *sawah*
Labor task (hours per hectare)				
Tractor rental	30	30	–	–
Animal plowing	–	–	110	150
Planting	200	200	200	200
Weeding and water management	516	516	416	230
Fertilizing/spraying	54	48	33	12
Paddy cutting	175	175	175	150
Manual threshing/winnowing	–	–	200	200
Machine threshing	40	40	–	–
Drying	24[a] 16[b]	24[a] 16[b]	24	24
Fertilizer (kiliograms per hectare)				
Urea	200	200	150	150
TSP	100	100	100	–
KCl	50	50	–	–
ZA	50	–	–	–
Other inputs				
Liquid insecticide (liters per hectare)	2	2	1	–
Granulated insecticide (kilograms per hectare)	8	8	–	–
Seeds (kilograms per hectare)	30	30	30	30
Spraying (times per hectare)	4	4	2	–
Yield (kilograms per hectare)	6,250	4,750	3,750	3,250

[a]Wet season.
[b]Dry season.

West Sumatra (rupiahs per unit)

Input	Private			Social		
	Tradable input	Domestic factor	Total	Tradable input	Domestic factor	Total
Rice seed (kilograms)	400		400	400		400
Urea (kilograms)	125		125	196	38	234
TSP (kilograms)	125		125	310	38	348
KCI (kilograms)	125		125	176	38	214
ZA (kilograms)	125		125	133	38	171
Granulated insecticide (kilograms)	700		700	1,352	27	1,379
Liquid insecticide (liters)	3,000		3,000	5,784	44	5,828
Sprayer rental (days)					850	850
Irrigation water (hectares)					6,500	6,500
Family labor (hours)		400	400		400	400
Animal services (hours)		1,150	1,150		1,150	1,150
Tractor services (hectares)	22,000	58,000	80,000	22,000	41,000	63,000
Machine thresher rental (hectares)		1/10	1/10		1/10	1/10
Traditional harvest (hectares)		1/9	1/9		1/9	1/9
Winnowing (hectares)		3%	3%		3%	3%
Planting labor (hours)		450	450		450	450
Transport (farm–mill) (kilograms)	0.25	4.75	5	0.25	3.8	4.05
Rice milling (kilograms)		10	10		10	10
Sacks (kilograms)	8		8	8		8
Transport (mill–market) (kilograms)	1	4	5	1	2.35	3.35

Java (rupiahs per unit)

Input	Private			Social		
	Tradable input	Domestic factor	Total	Tradable input	Domestic factor	Total
Rice seed (kilograms)	400		400	400		400
Soybean seed (kilograms)	1,000		1,000	1,000		1,000
Corn seed (kilograms)	1,000		1,000	1,000		1,000
Urea (kilograms)	120		120	194	30	224
TSP (kilograms)	120		120	308	30	338
KCI (kilograms)	120		120	174	30	204
ZA (kilograms)	120		120	131	30	161
Granulated insecticide (kilograms)	700		700	1,351	22	1,373
Liquid insecticide (liters)	3,000		3,000	5,783	37	5,820
Sprayer rental (days)					850	850
Irrigation water (hectares)					40,500	40,500
Male labor (hours)		237	237		237	237
Female labor (hours)		130	130		130	130
Planting labor (hours)		150	150		150	150
Family labor (hours)		215	215		215	215
Animal services (hours)		850	850		850	850
Harvest labor (hectares)		1/10	1/10		1/10	1/10
Transport (farm–mill) (kilograms)	0.25	4.75	5	0.25	3.8	4.05
Rice milling (kilograms)		10	10		10	10
Sacks (kilograms)	8		8	8		8
Transport (mill–market) (kilograms)	1	4	5	1	2.35	3.35

Java, sugarcane costs (ooo rupiahs per hectares

input	Private			Social		
	Tradable input	Domestic factor	Total	Tradable input	Domestic factor	Total
Preprocessing labor		800	800		800	800
Other preprocessing	295		295	375		375
Transport to mill	50	350	400	50	325	375
Processing/ marketing		1,231	1,231		1,159	1,159
Yield: 8,000 kilograms granulated sugar per hectare						

South Sulawesi (rupiahs per unit)

Input	Private			Social		
	Tradable input	Domestic factor	Total	Tradable input	Domestic factor	Total
Rice seed (kilograms)	300		300	300		300
Urea (kilograms)	125		125	196	38	234
TSP (kilograms)	125		125	310	38	348
KCI (kilograms)	125		125	176	38	214
ZA (kilograms)	125		125	133	38	171
Granulated insecticide (kilograms)	750		750	1,352	27	1,379
Liquid insecticide (liters)	3,000		3,000	5,784	44	5,828
Herbicide (liters)	6,000		6,000	11,556	44	11,600
Sprayer rental (days)					850	850
Irrigation water (hectares)					6,500	6,500
Family labor (hours)		250	250		250	250
Gotong-royong costs (hectares)		40,000	40,000		40,000	40,000
Animal services (hectares)		50,000	50,000		49,350	49,350
Tractor services (hectares)	20,000	45,000	65,000	20,000	33,000	53,000
Transport (farm–mill) (kilograms)	0.25	5.25	5.5	0.25	4.75	5
Rice milling (kilograms)		11	11		11	11
Sacks (kilograms)	8		8	8		8
Transport (mill–market) (kilograms)	1	14.5	15.5	1	4	5

8. Equity Effects of Rice Strategies

Rosamund Naylor

A more even distribution of income across groups of people and among regions of the country has long been a primary economic objective of the New Order government (Chapter 2). The importance of equity considerations in the selection of agricultural policies recently was highlighted in the Fifth Development Plan (REPELITA V) and in President Soeharto's budget speech of January 1990. This chapter provides an analysis of the equity implications of a trend self-sufficiency strategy for rice. At issue is whether the proposed package of incentive policies to promote increased rice production (discussed in Chapter 7) is likely to result in significant hardships for consumers of rice, thereby creating a trade-off between the government's food security and equity objectives.

The recent surge in rural incomes associated with successful economic growth policies has raised the income and employment levels of many of Indonesia's poorest people. But if future promotion of rice production will require increases in rice prices, some of the past improvements in equity may be offset. Because rice plays an important role in the expenditure patterns of poor and middle-income consumers, rice price increases will have a noticeable effect on the purchasing power of these groups. Following a strategy of rice self-sufficiency on trend thus may cause undesirable effects on personal income distribution.

Regional as well as personal distributions of income are relevant to policymakers. In this chapter, therefore, the analysis of income transfers associated with rice price increases considers rural and urban areas within five key provinces—East, Central, and West Java, South Sulawesi, and West Sumatra. Urban consumers are clear losers from price increases, but the effects on rural areas are less clear. As consumers, the rural population would prefer lower rice prices, but the negative impact of higher prices on rural rice consumers may be offset by the simultaneous effects of rice price increases on returns to labor and land.

Rice production is the largest employer of unskilled labor in the rural economy. Through their impact on the demand for labor, rice policies directly influence both the number of jobs and the wage rates offered to unskilled workers. Furthermore, effects on the labor market extend beyond those associated with direct employment of laborers in the paddy fields or rice mills. The substantial increases in incomes from rice during the past decade, realized as profits or higher returns to land, have allowed concomitant increases in expenditures by rice producers. Many of these expenditures—for construction, services, and locally produced commodities—augment the total demand for unskilled labor. Because these expenditure effects have been prominent in recent years, many observers of Indonesian agriculture characterize the rice sector as the "engine of growth" in the rural economy.

Linkages between Government Policies, Rice Production, and Rural Incomes

The equity effects of policies can be represented as a series of induced income flows to consumers, producers, and factors of production, illustrated in Figure 8.1. One category of equity effects is represented by the direct income transfers that result from rice price, crop regulation, and irrigation investment policies—the principal instruments available to policymakers in the design of a new rice strategy (Chapter 7). Solid lines indicate the direct income transfers from policies. Rice price policy affects the cost of living for consumers by altering the share of the budget that they spend on rice. In the event of a price increase, consumers usually will reduce the impact of the price change by substituting in favor of other staples in the food budget. But the consumer is still worse off than before the price increase. In contrast, price increases raise the incomes of rice producers because revenue increases translate into higher profits or returns to land. Producers also gain income from the changes in regulatory and investment policy considered here. A relaxation of cropping restrictions allows farmers to plant more profitable crops; in most areas, that decision favors increased production of rice. Investment policy increases incomes by allowing more intensive cropping of a given land area; that change usually augments the production of rice.

Another part of the equity effects of rice policies can be analyzed through an examination of wage and employment levels for unskilled labor. The economic welfare of hired laborers rises or falls with corresponding movements in the wage rate and aggregate demand for unskilled labor. The review of the literature and the fieldwork on the structure of the labor market (described in Chapter 5) provide evidence that the market for labor on Java is gradually becoming integrated. Because the costs of migration—between urban and rural areas and among rural areas—remain high, however, local demand and supply conditions have an important influence on wages. Wage rates, even for identical tasks, are not the same across regions. Relatively populated regions, such as Central

FIGURE 8.1. *Analyzing the equity effects of rice policies; solid arrows indicate direct effects, broken arrows indicate indirect effects*

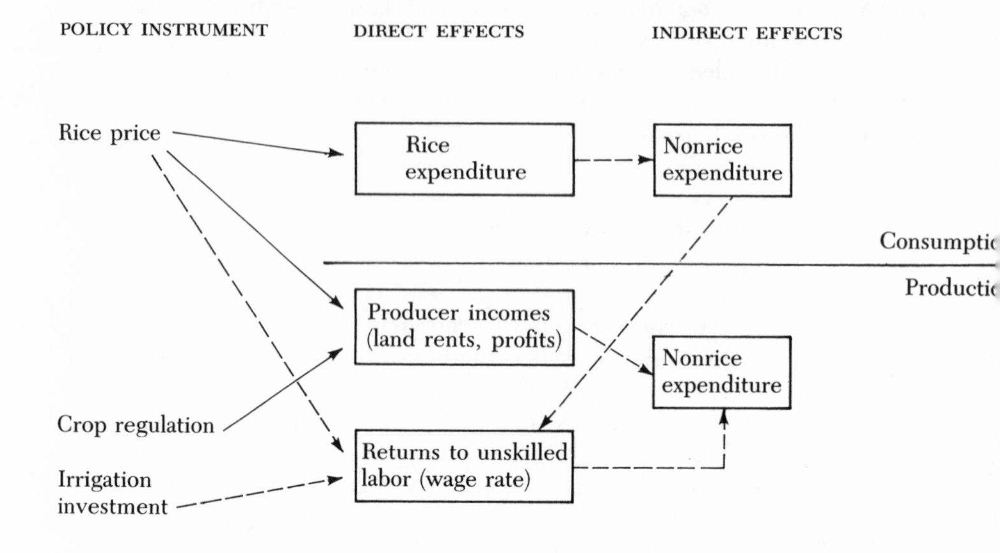

Java, have lower wage levels than the less heavily populated regions (Chapter 5).

The overwhelming impression, nonetheless, is one of an active and well-informed market for unskilled labor. Migration is widespread among rice laborers; nonrice employment for some part of the year is essential to generate an adequate annual income. Research on temporary and circular migration patterns confirms that mobility for rural households has increased dramatically since the mid-1970s because of the development of transportation and communication networks (Hugo 1981; Forbes 1981; Manning 1986; Sjahrir 1990). Workers now move easily to sectors and regions of the economy that offer employment opportunities.

Employment strategies—urban migration, migration to other agricultural areas, and working in local towns—differ among workers. Most workers are well aware of alternative wage rates, even in relatively isolated regions, and are ready to move to alternative employments if wage differentials become large enough. Wage rates vary among employments, but this wage structure usually can be explained by differences in the duration and difficulty of the job. Wages in each employment thus can be considered competitive with the alternatives.

Dotted lines in Figure 8.1 highlight the relationships relevant to the demand for unskilled labor. All of the policies considered in Figure 8.1 affect the market for labor—either directly, through the demand for labor needed to produce rice, or indirectly, through the impact on the level of nonrice expenditures by consumers and producers of rice.

A significant linkage ties price policy to the labor market. Increases in rice prices raise the marginal value product of labor used in rice production. Initially, the value of these productivity increases will benefit producers in the form of higher profits. But if labor markets are competitive, increases in profitability will be passed on to workers through increased employment, higher wages, or some combination of the two. Increases in rice prices encourage producers to use more labor by planting more area to rice and by more careful planting, tending, and harvesting of the existing area. When producers compete for the services of hired labor, wages will have to increase to induce the necessary workers away from their alternative employments. If rice is unimportant in the total demand for labor, wage increases will be trivial; the necessary labor could be found easily without offering wages that are very different from prevailing rates. But if rice plays a large role in the total employment of unskilled workers, the impact of rice price increases on wage levels becomes more important.

The linkages between crop regulatory and irrigation investment policies and the labor market depend on labor use in rice production relative to that needed for alternative crops. Irrigation investment augments total agricultural area by creating more opportunities for multiple cropping. Rice is expected to be the main beneficiary of this investment; because rice is the most labor-intensive primary food crop, the total demand for labor will be increased. A relaxation in regulatory policy also will increase the aggregate demand for labor because rice production is more labor-intensive than the crops (sugar and tobacco) that are now grown on the regulated area.

The indirect effects of rice policies on the labor market are more difficult to trace. A change in profits accruing to rice farmers adds to household income that is either spent on the consumption of food and nonfood items, saved, or invested. Similar income flows are generated by a change in wage income in the rice sector. These expenditures and investments contribute indirectly to the growth in employment opportunities for unskilled labor, as long as the goods and services demanded are produced domestically (Hazell and Roell 1983). The extent to which expenditures create employment opportunities in rural areas depends on the predominant expenditures (e.g., housing, services, or education) and on the locations of production for the goods and services demanded (e.g., rural or urban).

The relevance of expenditure effects for rural wage rates and employment is also contingent upon the policies that initiate the flow of incomes between the product and factor markets. Public investments in technology, such as high-yielding seed varieties, and investments in irrigation and marketing infrastructure generally have positive impacts on rural incomes and employment. But these gains can be offset if the investment funds are raised through taxes on producers or consumers in rural areas. Rice price policy also has an ambiguous indirect effect because the expenditure of increased profits from rice production will be offset at least partially by a decline in real incomes for net consumers of rice.

Income Transfers from Rice Policies

Real incomes from rice farming more than doubled during the past twenty years (Chapter 6). Rice price and investment policies—including irrigation expansion, subsidies on farm inputs, and a guaranteed floor price for output—have contributed to much higher farm incomes. The impact of these policies on income distribution, however, remains a moot issue. This section investigates the income effects of the three policies proposed to implement the strategy of rice self-sufficiency on trend.

Rice Price Policy

A 5 percent increase in the real price of rice would generate transfers of income from urban to rural areas and from consumers to producers within the rural economy. The magnitudes of these transfers are determined by the amount of rice produced and consumed by each income group and the responses of supply and demand to the rice price change. The income distribution of individuals in rural and urban areas by region, as reported in the 1984 SUSENAS survey, is shown in Table 8.1.[1] Two distributions are shown, one for consumers and the other for producers. The classification of consumers is the most comprehensive summary of the income distribution because the sample consists of all residents in each region. The income distribution statistics for rural producers refer to the subset of the population that own or lease land.[2]

[1] The income distribution represents the proportion of the population in each expenditure class. The terms *income distribution* and *distribution by expenditure class* are used interchangeably in the text, assuming that the savings rate is roughly equivalent among income classes. The expenditure classification shown in Table 8.1 represents a distribution around the middle-income group (almost 60 percent of the total Indonesian population), with skewness toward the lower-income group (32 percent of the total population). Less than 10 percent of the total population falls in the high-income group.

[2] The income distribution of producers was calculated from rice consumption data reported in the 1984 SUSENAS survey. Data on rice consumption in the SUSENAS survey are disaggregated into purchased and own-produced rice. The quantity of own-produced rice in each income class as a proportion of total own-produced rice (for all income classes combined) was used to infer the percentage of producers in each income group.

TABLE 8.1. *Income distribution by region*

	Consumers[a]			Producers[a]		
	Low	Middle	High	Low	Middle	High
Rural areas						
West Java	25.5	69.6	4.9	27.6	36.9	35.5
Central Java	60.2	37.9	1.9	23.8	37.8	38.4
East Java	49.5	47.4	3.1	20.1	36.7	43.2
South Sulawesi	44.2	49.8	6.0	29.1	44.0	26.9
West Sumatra	13.2	78.9	7.9	28.7	37.4	34.0
Indonesia AVERAGE	39.4	56.6	4.0	23.4	39.7	36.8
Urban Areas						
West Java	9.8	67.4	22.8			
Central Java	18.3	69.7	12.0			
East Java	12.9	72.6	14.5			
South Sulawesi	10.9	70.7	18.4			
West Sumatra	2.0	63.5	34.5			
Indonesia AVERAGE	8.8	66.8	24.4			

SOURCE: BPS, SUSENAS, 1984 survey.
NOTE: Per capita monthly expenditure class (1984): low: <Rp 10,000; medium: Rp 10,000–30,000; high: >Rp 30,000.
[a]Percentage of population in each category.

The distribution of the consumer population by income group varies between urban and rural areas. Within urban areas, about two-thirds of consumers are in the middle-expenditure category. About one-fourth of the average urban population is in the high-income category, although this proportion varies by province. In comparison, rural areas have a much higher incidence of relative poverty. Except for West Sumatra, at least one-fourth of the rural population in each province studied is in the low-income category. This proportion is as large as 60 percent (Central Java).

The income distribution of rice producers (who are assumed to live in rural areas only) is spread more evenly and across all income groups, with some bias toward the middle- and higher-expenditure categories. About three-fourths of producers are in the middle- and upper-income categories. In all provinces, except South Sulawesi, more than one-third of rice producers are in the highest expenditure group. Given this distribution, the direct income transfers from rice policies will favor relatively wealthy households more than poor households.

Table 8.2 shows the short-run income transfers by region resulting from a 5 percent increase in rice prices. These transfers reflect producers' surplus gains and consumers' income losses starting with a base period assumption of national rice self-sufficiency. The gains and losses are calculated from regional production and consumption data and aggregate

TABLE 8.2. *Income effects of price policy*

Net transfer of income between consumers and producers with a 5 percent increase (million rupiahs, constant 1988 prices)[a]

Region/country	Rural			Urban		
	Low	Middle	High	Low	Middle	High
West Java	11,479	−16,911	32,559	−1,597	−11,111	−3,926
Central Java	−3,117	10,171	21,421	−1,684	−8,492	−1,647
East Java	946	18,311	35,966	−1,420	−9,227	−2,174
South Sulawesi	3,920	4.767	7,390	−413	−2,515	−595
West Sumatra	4,191	−1,756	5,473	−201	−704	−533
Indonesia	−173	−4,599	153,513	−6,116	−60,190	−26,465

Consumer surplus losses per capita (rupiahs per capita)

Region/country	Rural			Urban		
	Low	Middle	High	Low	Middle	High
West Java	−2,544	−3,028	−3,245	−2,119	−2,357	−2,394
Central Java	−1,534	−2,229	−2,667	−1,643	−2,021	−2,185
East Java	−1,232	−1,957	−2,121	−1,534	−1,957	−2,121
South Sulawesi	−1,961	−3,158	−3,454	−1,917	−2,454	−2,763
West Sumatra	−1,851	−2,947	−3,004	−1,871	−2,180	−2,474
Indonesia	−2,028	−2,452	−2,924	−1,755	−2,165	−2,635

Per capita consumer losses as percent of annual per capita expenditures[b]

Region/country	Rural			Urban		
	Low	Middle	High	Low	Middle	High
West Java	1.53	0.79	0.22	1.25	0.61	0.16
Central Java	1.06	0.64	0.29	1.09	0.53	0.17
East Java	0.52	0.43	0.12	0.78	0.46	0.16
South Sulawesi	1.12	0.68	0.21	0.87	0.51	0.20
West Sumatra	1.00	0.86	0.24	0.74	0.66	0.27
Indonesia	0.92	0.70	0.21	1.01	0.55	0.17

SOURCE: Author's calculations, detailed in Appendix 8.1.
[a]Government budgetary implications are not considered in these calculations.
[b]Expenditure shares are derived from the 1984 SUSENAS data.

short-run price elasticities of supply and demand.[3] Price elasticities of demand incorporate substitution and regional income effects. (Assumptions and detailed results for the analysis of income transfers are presented in Appendix 8.1.)

A 5 percent increase in the rice price would transfer income from urban to rural areas and from low- and middle-income groups to the high-income group within rural regions, as shown in part 1 of the table. The largest gains from the price increase accrue to the high-income class in

[3] A 5 percent price increase for *gabah* is translated into a 3.6 percent increase in the retail price, by assuming a 65 percent milling ratio and a constant marketing margin. The results are highly sensitive to marketing assumptions; for further reference on rice marketing in Indonesia, see Mears (1981).

rural areas, a group that encompasses more than one-third of rice producers but only 4 percent of the rural population. Aggregate income losses would be greatest for the middle-income category in urban locations, reflecting its large population share. Although many rural producing regions would benefit from the price rise, the low-income group in Central Java and the middle-income groups in West Java and West Sumatra—both with relatively large populations—would suffer net losses. This result challenges the widely held presumption that income transfers from urban to rural areas necessarily have positive equity effects in a rural-based economy.

The implications of the rice price increase for per capita expenditure are summarized in the lower third of Table 8.2. These data show the regressive incidence of rice price policy on consumers. Even though members of the low-income cohort consume less rice than people in the medium- and high-income categories, the welfare losses are greatest for the poorest group. Relative to per capita income, consumers in the low-income group lose half again as much as middle-income consumers and about four times as much as high-income consumers. However, the absolute magnitudes of income losses are small. Losses are about 1 percent or less of low-income consumer expenditures, except for West Java (1.25 to 1.53 percent). For the medium- and high-income classes, losses are generally less than 0.8 percent of expenditure.

Irrigation Investments

Investments in new irrigation systems or the rehabilitation of current irrigation systems would have a favorable impact on producers' incomes without penalizing consumers, unless the investment funds were raised through new taxes.[4] The principal income transfers involve the distribution of income gains between regions and among income groups within regions. The regional distribution of benefits depends on the location of planned projects, giving policymakers the capability to target areas of future production, income, and employment growth.

Within regions receiving investment funds, the size of the income transfer is affected by the nature of the project (e.g., rehabilitation of existing systems or conversion to production systems with better water control). Investment funds spent on the rehabilitation of technical irrigation systems in high-productivity areas on Java would have markedly different distributional consequences relative to funds spent on new land development in poorer regions off Java.

[4] It is assumed that the investment funds are raised through the taxation of rents unrelated to the rural sector of the economy. Investments in irrigation have opportunity costs regardless of the source of investment funds, however. The opportunity costs associated with public sector investments in infrastructure development warrant further investigation.

Estimates of additional income to producers resulting from irrigation expansion and rehabilitation under the proposed strategy are contained in Table 8.3. The calculations are based on profitability per hectare for each type of investment multiplied by the incremental cropping intensity and number of hectares targeted for improvement in REPELITA V (Table 7.4).

The development of new irrigated land and the completion of existing projects on new land would generate the greatest increase in profitabilities. Relatively large income gains for these categories result from the substantial planned expansion of irrigated land, which amounts to almost 3 percent of existing rice area. These investments would direct income to farmers currently without high-productivity land, both on and off Java. A much smaller share of the funds would accrue to farmers in swampland regions, many of whom are in the low-income group.

Planned expenditures on conventional rehabilitation and the conversion of rainfed to irrigated systems are equally divided between Java and off Java. These investments would benefit farmers in all income groups. More information on specific projects within the plan is needed before a detailed assessment of income transfers is possible. But the data contained in Tables 7.4 and 8.3 indicate that the transfers would not favor only higher-income farmers owning good-control *sawah*.

Sugar Policy

A gradual reduction in the mandatory number of hectares devoted to sugar production offers an immediate means to increase rice production and improve farmers' incomes. Given the location of land under the TRI

TABLE 8.3. *Additional income generated by irrigation expansion and rehabilitation during REPELITA V (billion rupiahs, constant 1988 prices)*

Category of investment	Total income gain
Upgrading of swampland	8.0
Conventional rehabilitation	
Java	10.5
Off Java	5.2
Extensification	
Existing projects	105.0
New land development	81.3
Conversion of rainfed land to *sawah*	
Java	4.0
Off Java	9.5
TOTAL	223.5

SOURCE: Author's calculations from irrigation estimates in Table 7.4 and from farm budgets compiled during field surveys.

TABLE 8.4. *Cumulative income growth to rice producers from a phaseout of TRI policy by 1994 (billion rupiahs, constant 1988 prices)*

Income Class[a]	East Java	Central Java	West Java	Total
Low	17.9	10.8	2.0	30.7
Middle	32.7	17.1	2.7	52.5
High	38.5	17.4	2.5	58.4
TOTAL	89.1	45.3	7.2	141.6

SOURCE: Author's calculations from field survey data.
[a]Monthly expenditures per capita: low: <Rp 10,000; medium: Rp 10,000–30,000; high: >Rp 30,000.

program, the proposed phaseout of sugar primarily would benefit farmers who own medium- and good-control *sawah* in East and Central Java. Table 8.4 shows the additional income in constant 1988 prices that could be earned by producers under a phaseout of the TRI program.

The largest share of the income gains in East Java would accrue to producers in the high-income group. Additional income in Central and West Java would be split more evenly between the middle- and high-income groups. Over three-quarters of the total increase would go to middle- and high-income producers. Income growth in Central Java associated with the proposed phaseout would help to offset losses for the low-income group arising from the increase in the rice price.

Total Transfers

The transfer effects of rice price, irrigation, and sugar policies, forming a strategy to maintain rice self-sufficiency on trend, primarily would benefit rice producers, a high-income group. The exact pattern of regional income distribution cannot be calculated without more information on the location of planned irrigation projects. But the aggregate change in income levels within the rice sector can be predicted. The income transfer effects of the policies for the 1989–93 planning period are summarized in Table 8.5. The largest net gains in income would result from irrigation

TABLE 8.5. *Net income resulting from the proposed rice strategy (billion rupiahs, constant 1988 prices)*

Policy	Total income	Income per million additional tons produced
Price policy	56.0	16.0
TRI phaseout	141.6	88.1
Irrigation investments	223.5	113.5
TOTAL	421.1	217.6

SOURCE: Tables 7.6, 8.2, 8.3, and 8.4.

investments. Net income growth from the combined set of policies, Rp 421 billion, is equivalent to 10 percent of total income earned in rice farming on Java in 1987 (Chapter 6).

Rice Policies and the Labor Market

The additional income generated by the proposed strategy would be divided among the factors of production within the rice economy. Landowners would be expected to gain from increased returns to land and management. Increases in land rental values would benefit primarily the higher-income, landowning individuals in rural areas. Although the population of producers is well represented in each of the three income categories (Table 8.1), the low-income category of farmers has smaller landholdings. Land rental is also more important among this group, at least in relative terms, and the benefits of higher returns to land go to owners rather than tenants (Manning and Wiradi 1984). Landless laborers and family labor also would gain if they receive higher wages associated with future growth in total demand for unskilled labor. Wage increases would benefit all income categories; the aggregate benefits would be proportional to total individual employment.

Rice and Employment

The magnitude of wage rate effects in the unskilled labor market depends on the importance of rice in the total demand for labor. Aggregate and microeconomic evidence shows that the rice sector is becoming relatively less important over time as a source of employment. Rice employment has remained almost constant since 1969. Technical change, especially the introduction of sickle harvesting, has allowed per hectare demands for labor to decline; these decreases have been offset by increases in cultivated area. In 1969, for example, an estimated 7 billion labor hours were required to produce 11 million tons of rice, whereas in 1987 approximately the same labor input was able to grow 24.5 million tons of rice. The rest of the agricultural sector appears to have behaved similarly, although employment data are incomplete and not fully reliable. Aggregate employment statistics for Java show annual employment growth rates in agriculture of less than 1 percent, whereas total employment grew at about 2.5 percent per year (Naylor 1989, pp. 154–156).

But in spite of its diminished role, rice still accounts for a prominent part of total employment. Although exact estimates of the proportion of employment provided by rice are not available, manipulations of existing data can provide a rough idea of the magnitude. Aggregate statistics indicate that the agricultural sector provides as much as two-thirds of the total employment of unskilled labor.[5] These statistics do not allocate agricultural labor among commodities. Commodity-specific statistics are available for the shares of rice in agricultural GDP (28 percent) and in harvested area (about 40 percent); these proportions understate the importance of rice in agricultural employment because rice production is more labor-intensive than most crops. If rice were to account for as much as half of agricultural employment, the share of rice in total employment of unskilled labor would be about one-third.

Approaching the estimation problem from a more microeconomic perspective points to a lower proportion for the share of rice in total employment. The production budgets developed for rice provide estimates of labor inputs per hectare (Chapter 4); these figures can be combined with data on total cropped area to yield an estimate of the total labor use in rice production. If it is assumed that 150 days represent the average annual employment for an unskilled laborer, total labor demand can be converted into years of employment. This number is then compared with the size of the working population to generate the share of rice in total employment. This procedure yields an estimate of 22 percent for the employment share of rice. The share varies among regions; it is relatively large for West Java (27 percent) and relatively small for Central Java (16 percent) (Naylor 1989, pp. 160–61). These results are consistent with the pattern of employment opportunities described in Chapter 5.

The prominence of rice in the labor market—between 20 and 30 percent of unskilled employment—is likely to continue into the next decade, particularly if policymakers choose a strategy to promote rice self-sufficiency on trend. The set of policies described in the previous chapter could create almost 1 million additional jobs in rice production by the end of 1993.[6] Table 8.6 shows the estimated gains in agricultural employment resulting from area responses to price policy, the shift of TRI sugar land into rice production, and increased cropping intensities and area extension from planned irrigation investment. The measures are overestimated because the growth in on-farm employment from price policy—116 full-time jobs for each additional thousand tons of *gabah* produced—would be

[5] Recent statistics reported by BPS show that the agricultural sector had about 55 percent of total employment in 1986 (Naylor 1989, p. 10). If it is assumed that 15 percent of the total labor force is skilled labor and that the entire agricultural labor force can be classified as unskilled, the share of agriculture in unskilled employment can be estimated at 65 percent.

[6] These calculations are based on an assumption of no labor input substitution in rice production. Full employment in the rice sector is estimated at 180 days per year on Java, 100 days per year off Java, and 150 days per year for Indonesia as a whole. The average workday in rice production is assumed to be 5 hours. Labor input coefficients used in the calculations are based on 1988 estimates of 1,460 labor hours per hectare on Java, 550 labor hours per hectare off Java, and 1,150 labor hours per hectare for Indonesia as a whole (field survey data).

TABLE 8.6. *Net gains in rice employment by 1993*

	Sources of production growth			
	Price and trend[a]	Sugar policy	Irrigation investments	Total employment
Total labor years	406,333	150,000	570,400	1,126,733
Labor years per thousand tons of additional *gabah*	116	93	290	159

SOURCE: Author's calculations, based on projections of area and production growth given in Table 7.6.
[a]Does not include negative cross-price effects on alternative crops and possible labor-saving technical changes.

offset partially by employment losses from the reduced area planted in substitute crops. Moreover, labor-saving technical changes, like those that took place during the past two decades, could offset some or all of these potential new jobs in rice production.

The largest increase in rice labor demand is expected to come from planned irrigation investments. These investments will bolster agricultural employment in the long run through increased cropping intensities and area expansion. In addition, unskilled labor will be required for the construction and rehabilitation of irrigation facilities. Because of the indivisible nature of irrigation projects, employment growth associated with irrigation investments would be concentrated in the locations of new or improved facilities. These locations are not specified precisely in REPELITA V (Varley 1989). But most of the growth in area is expected to come from the completion of existing projects and new land development. As reported in Chapter 7, irrigation extensification will apply increasingly to remote areas because many accessible areas already have been developed. This pattern of development would improve the welfare of landless laborers in remote regions, who are often in need of additional employment opportunities because relatively few off-farm activities exist nearby.

Net employment growth resulting from a phaseout of the national sugar program also would have a regional impact on demand for labor. Impacts of this policy change would be concentrated in East and Central Java. Of the 150,000 hectares of *sawah* that could be added to rice production by the end of 1993 under a reform of the current policy, roughly two-thirds are in East Java and one-third are in Central Java. A very small proportion (less than 5 percent) is in West Java. Estimates of additional demand for labor in agriculture from the phaseout plan are shown in Table 8.7.

A phaseout of the sugar program would allow *sawah* that is currently planted in sugar for sixteen months to be planted with as many as three rice crops and one or more *palawija* crops. The on-farm employment opportunities for various cropping patterns of rice and *palawija* greatly outweigh the labor requirements for a single sugar crop. Sugar processing uses labor between the months of May and December when the sugar is milled, but the employment provided by processing does not compensate for sugar's lower on-farm labor requirements.

The employment effects of the set of policies associated with trend self-sufficiency could create enough demand for labor so that total labor use in rice production will not fall during the next five years, even with substantial input substitution for labor. In the absence of any labor substitution, employment gains in rice farming would be equivalent to roughly 15 percent of current rice employment, between 3 and 5 percent of the total employment for unskilled labor. The prospects seem good that rice employment can retain its recent labor share, thus maintaining the prominence of rice in the labor market.

Rice and the Wage Rate

Real wages rose during the first half of the 1980s throughout Java (Chapter 5). Although this growth coincided with rising productivity and increasing prices in the rice sector, increased demand for labor in nonagricultural activities in both rural and urban locations was crucial. The importance for the labor market of demand forces external to the rice sector is suggested by comparisons of changes in wages with changes in land rental rates. Except for land subject to demand for conversion into nonagricultural uses, changes in the sales prices and rental rates for land are influenced principally by changes in the profitability of rice production. Net returns to fixed factors in rice production (rice-farming income) rose by over 50 percent during the first half of the 1980s (Table 6.7).

Table 8.8 shows the ratios of land rental prices to wages in rice production. Such comparisons are imprecise because of lags in the response of factor prices to changes in the demand conditions for land and labor (Jatileksono 1989). But the general trends suggest that the ratios of land

TABLE 8.7. *Additional employment under a sugar phaseout policy, 1989–94 (000 labor years, cumulative)*

	East Java	Central Java	West Java	Total
1989	5	2	0.4	7.4
1990	35	18	3	56
1991	54	27	4	85
1992	73	37	6	116
1993	91	46	7	144
1994	94	48	8	150

SOURCE: Author's calculations, based on labor coefficients from field surveys.
NOTE: Includes jobs generated in rice and *palawija* production and processing made possible by phasing out sugar.

TABLE 8.8. *Ratios of land rental prices for* sawah *to wages in rice production, 1976–86*

	Ratio of land rental to hoeing wages			Ratio of land rental to weeding wages		
	West Java	Central Java	East Java	West Java	Central Java	East Java
1976	166	721	523	279	828	682
1977	151	598	557	267	704	712
1978	161	564	566	280	662	764
1979	217	508	564	374	614	773
1980	193	446	502	327	535	655
1981	200	399	462	329	502	589
1982	183	418	403	298	519	515
1983	168	373	338	263	452	468
1984	168	323	323	266	393	443
1985	213	300	323	347	361	472
1986	204	291	321	331	349	478

SOURCE: BPS, "Farmers' Terms of Trade Index."

rentals to wage rates rose slightly in West Java (by perhaps 20 percent) and fell substantially in Central and East Java (between 40 and 50 percent). This result is consistent with evidence on real wage trends presented in Chapter 5, which showed much stronger growth in real wages (and off-farm employment opportunities) in East and Central Java than in West Java. Although the absolute ratios of *sawah* rental prices to wages were much higher in Central and East Java than in West Java, the ratios have gradually converged in recent years. Hence the welfare of laborers within the rice sector improved in relation to that of landowners, at least in measurable factor returns.

Available data do not permit a precise specification of the role of the rice sector in influencing the wage rate for unskilled labor. But the importance of rice in total employment makes it difficult to conclude that recent growth rates in wages would have occurred without substantial increases in rice prices and productivities. Productivity growth from technical change and higher rice prices increases the ability of producers to pay for labor services and compete against alternative employment opportunities. Because the rice sector is large absolutely, changes in that sector's demand create potential outflows of labor from it or inflows to it that are relatively large as well. The consequent impact on aggregate labor productivity and market wage rates should be noticeable. The derivation of the transmission effect is summarized in Appendix 8.2.

Since estimates of labor demand elasticities are not available, it is possible to venture only very rough guesses about the prospective magnitude of the transmission effect. The proposed self-sufficiency strategy foresees a rice price increase of 5 percent in 1991. If it is assumed arbitrarily that the labor demand elasticities in the rice and nonrice sectors are equal, the impact of that rice price hike on rural wage rates can be approximated by multiplying the percentage price rise by the ratio of the share of rice in total rural unskilled employment to the share of nonrice in that employment. For example, if rice has a 30 percent share of unskilled employment, the impact on the unskilled wage rate of a 5 percent rise in the rice price could be about 2 percent (0.3/0.7 times 5 percent). Alternatively, if the share of rice in total unskilled employment is 20 percent, the estimated transmission effect would be 1.25 percent (0.2/0.8 times 5 percent). These calculations illustrate that the transmission effect could be significant, perhaps as high as 1 or 2 percent of the wage rate. This linkage would provide a potentially important favorable offset to the negative impact of higher rice prices on low-income consumers who are unskilled laborers in either rural or urban areas.

Rice and Expenditure Effects

Past expenditures and investments of households benefiting from increased returns to land and labor have led to higher levels of demand for off-farm goods and services. That increased demand, in turn, has generated demands for unskilled labor in a variety of rural nonagricultural activities. These changes have contributed to the growth in total demand for unskilled labor in the economy and, therefore, are partially responsible for the increases in real wage rates since 1976. But the impacts of the policy initiatives under consideration here are less clear. Net rice consumers will be forced to reduce expenditures on commodities other than rice, whereas net producers will be able to increase nonrice expenditures. Whether this transfer of purchasing power will benefit unskilled laborers depends on the expenditure patterns of the initial gainers and losers from price policy.

Data on budget propensities by expenditure group and region, contained in the SUSENAS survey (1984), are presented in Table 8.9. Household expenditure patterns by income class vary considerably. The share of household income spent on basic needs (food, housing, and clothing) is largest for the low-income group and declines as income levels rise. A sizable proportion of the household budget in the high-income class is allocated to durable goods, such as furniture, vehicles, and jewelry, as well as to education, health, and domestic servants.

In rural areas, roughly two-thirds of household income is devoted to the consumption of food items, and the remaining one-third is spent on nonagricultural goods and services, principally housing and utilities. Rice is the main expenditure item within the food category, accounting for 20 to 26 percent of total expenditures by the low- and middle-income groups,

TABLE 8.9. *Average per capita monthly expenditure by income class (percent of total monthly expenditures, 1984)*

Expenditure item	Consumers[a]			Producers[a]		
	Low	Middle	High	Low	Middle	High
Total food	70.5	67.1	39.8	68.6	59.0	39.0
Rice	26.0	19.7	5.9	28.6	15.4	4.9
Total nonfood	29.5	32.9	60.2	31.4	41.0	61.0
Housing and utilities	17.0	15.1	19.0	19.4	21.0	27.1
Clothing and footwear	4.8	5.0	3.5	4.8	4.9	3.9
Durable goods	1.1	3.2	18.8	0.6	1.6	6.1
Parties and ceremonies	2.8	3.2	4.9	1.7	1.7	2.4
Miscellaneous goods and services[a]	3.2	5.8	12.2	4.6	10.8	19.3
Taxes and insurance	0.5	0.7	1.8	0.3	0.9	2.2
Total food and nonfood	100.0	100.0	100.0	100.0	100.0	100.0

SOURCE: BPS, SUSENAS, 1984 survey.

NOTE: Per capita monthly expenditure classes (1984): low: <Rp 10,000; middle: Rp 10,000–30,000; high >Rp 30,000.

[a]Includes health, education, transportation expenses, and domestic servants.

which together constitute 92 percent of the rural population. For these income groups, housing and utilities are the dominant nonfood expenditures, followed by clothing and miscellaneous goods and services (e.g., health, education, and transportation expenses). Expenditures on parties and ceremonies, such as weddings and religious festivals, constitute roughly 10 percent of nonagricultural spending for all income classes.

A similar expenditure pattern exists for urban households. Food expenditures account for a slightly smaller share of the household budget in urban areas than in rural areas, although the share is still between 40 and 70 percent. Rice is the principal food expenditure for the lower- and middle-income groups. Within the nonfood category, the proportion of household budgets spent on housing is larger for urban areas than for rural areas, whereas the opposite is true for expenditures on durable goods and ceremonies. Urban households spend a significant portion of their incomes on education, health, domestic servants, and transportation expenses.

In most developing countries, the income elasticity of demand by rural households for nonfood consumption items is positive and often exceeds unity (Mellor 1984; Monteverde 1987; Timmer et al. 1983). Examples of rural-based goods and services with high income elasticities of demand are housing, education, and transportation. Accordingly, an increase in income received by relatively wealthy Indonesian rice producers, as a result of changed rice and sugar policies, would raise the demand for these goods and services in rural areas.

Labor intensities in the production of goods and services in each expenditure category underlie the potential employment effects of policy-induced income transfers. Within the nonfood category, the building industry offers the most widespread employment opportunities, particularly for men (Chapter 5). Investments and expenditures on housing provide jobs in construction and in the manufacturing of construction materials such as bricks, plywood, tiles, and cement. In many of the villages surveyed, both on and off Java, men find local jobs in the dry season within the construction sector and do not have to leave their village or region to work. In villages where profits from rice production are particularly important, men migrate to large urban centers less frequently than in earlier years because of an increase in local construction activity. An increment in housing demand resulting from rice policies, therefore, would stimulate labor demand.

The clothing and textile industries are labor-intensive and employ unskilled labor on a seasonal and full-time basis at wages comparable to those in the agricultural sector. But these industries are not expected to benefit much from the proposed rice strategy because of the low budget share of clothing, particularly for high-income households. Furthermore, short-run income losses for many poor households, resulting from price policy, would curtail the demand for low-cost services (e.g., petty trade) that are often provided by unskilled labor. An anticipated rise in spending on transportation and domestic servants by the high-income group, however, would create additional demand for local unskilled labor.

The demand for labor in the agricultural sector also would be affected by changing patterns of food demand. Consumer demand for staples such as rice that have relatively low income elasticities of demand would change less than that for products with high income elasticities of demand such as livestock, fruits, and vegetables. These labor-intensive commodities offer a range of employment opportunities in production and processing. The expected increase in incomes resulting from rice policies in the long run, therefore, would enhance labor demand in both agricultural and nonagricultural activities.

Expenditure patterns derived from budget propensities, however, do not necessarily reflect the marginal changes in the patterns of consumption as incomes rise. For example, field observations indicate that education is becoming increasingly important for rural society in Indonesia. Rising expenditures on education most likely will improve human capital in both rural and urban areas but may not provide many jobs for unskilled labor in the short run. In this case, the existing data lead to an overestimate of the indirect effects of rice policies on the wage rate and employment.

Although the pattern of income transfers from rice policies seems likely to lead to increased demand for labor, this assessment of indirect employment growth resulting from rice policies cannot be conclusive. Data are insufficient to indicate anything other than plausible directions of change in total demand. The results discussed here are similar to those of other studies; income transfers favoring those with relatively high incomes have the effect of increasing the demand for unskilled labor (Hazell and Roell 1983). But the magnitude of that effect for Indonesia remains unknown.

Conclusion

An emerging trade-off between the objectives of food security and efficient income growth is likely to be encountered in pursuing a strategy of rice self-sufficiency on trend (Chapter 7). But the analysis of linkages between that rice strategy and the unskilled labor market shows that any future trade-offs with income distributional objectives probably will not be large. Whereas the direct income transfers from rice policies would accrue primarily to relatively high-income households in the rural areas, the evidence considered in this chapter also provides plausible grounds for expecting significant offsets to the negative impact of higher rice prices on consumers.

Higher rice prices transfer income from urban to rural areas and reduce the purchasing power of net consumers of rice in all regions. But this reduction in purchasing power would be offset partially by the impetus given to demand for unskilled labor. Rice is the most significant employer of unskilled labor; even though its relative importance has diminished substantially as the economy has developed, rice production continues to play a key role in wage rate formation. Rice price policy thus should benefit unskilled labor—directly, through the transmission of rice price increases to wage rates, and indirectly, through the demands of rice producers for goods and services that require substantial amounts of unskilled labor in their production. Although the directions of these offsetting effects are clear, their magnitudes are not measurable with existing information.

The proposed changes in regulatory and investment policies, necessary for the attainment of rice self-sufficiency on trend, would help maintain the prominent role for rice in unskilled labor employment. Employment in rice production could increase as a result of the proposed strategy; in contrast, the number of jobs in rice farming and milling probably would fall without further production incentives. The magnitude of the changes in the wage rate from regulatory and investment policies cannot be estimated. In addition, further research on savings, investments, and expenditures of farm households is required before a definitive assessment can be made of the cumulative effects of rice policies on the rural labor market. But the positive effects of higher wages and more jobs can be expected to provide some offset to the negative effect of higher rice prices on consumption expenditures of the poor.

One of the principal influences on the future equity effects of any rice strategy will be the growth rate of the Indonesian economy. If national income growth rates remain high, real wages should continue to increase. Rising real wages, associated with strong demand for labor outside of the rice sector, would have two counteracting effects on equity in the rural areas. By increasing incomes for unskilled laborers, rising real wage rates would ameliorate adjustments to higher consumption costs of rice. But the growing labor costs also would be likely to trigger greater substitution of purchased inputs for labor, thus reducing labor absorption in rice agriculture. These changes would diminish the role of rice in total employment and weaken the transmission of rice price changes and rice-dependent expenditure effects. But because such changes would be triggered by rising wage rates, the equity impacts of rice policy would be easier to tolerate—for policymakers and consumers.

Appendix 8.1. Analysis of Income Transfers

This appendix contains the data, assumptions, and results of the analysis of income transfers between consumers and producers. Income transfers are calculated on the basis of the change in consumers' incomes (a) and the change in producers' surplus (b) for gross (not net) producers and consumers of rice as a result of a change in the rice price.

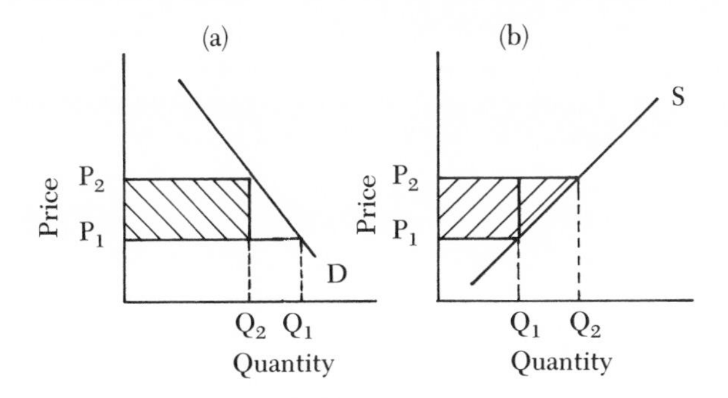

The short-run gains to producers and losses to consumers are calculated as follows, assuming linear demand and supply curves around the price change:

Consumers' income loss: $(P_1 - P_2) * Q_2$
Producers' surplus gain: $(P_2 - P_1) * Q_1 + 0.5 * (P_2 - P_1) * (Q_2 - Q_1)$

The first section of results contains the information on changes in consumers' income. The price elasticity of demand for rice is calculated with the income elasticity and the compensated demand elasticity using the Slutsky formula. A single compensated demand elasticity is used for all income groups. The Slutsky equation is:

$$\epsilon_{11} = \xi_{11} - b\eta_1$$

where ϵ_{11} = the own-price elasticity of demand for rice,
ξ_{11} = the compensated elasticity of demand for rice,
b = the budget share spent on rice, and
η_1 = the income elasticity for rice.

The short-run elasticity assumptions are as follows (Tabor et al. 1988):

	Income group		
	Low	Middle	High
Income elasticity:	0.7	0.4	0.2
Compensated demand elasticity: -0.14			

The average per capita consumption of rice in the base period (1988) is assumed to be 143 kilograms per year. This estimate is consistent with the calculation of consumption equivalent production for 1988 using a population estimate of 176 million. The average consumption level is distributed by region and income group according to the 1984 SUSENAS survey.

The second section of results shows the change in producers' surplus by region. The elasticity of supply with respect to the real rice price is 0.2 in the short run, and the initial production level for 1988 is 43.2 million tons (Table 7.6).

Net income transfers among consumers and producers are shown in the final table. Assumptions for initial (average 1988) paddy and rice prices are Rp 210 and Rp 450 per kilogram, respectively. Constant 1988 prices are used throughout the analysis.

TABLE A8.1. *Summary table on income transfers*

Change in consumers' surplus (million rupiahs)

	Rural			Urban		
Region/country	Low	Middle	High	Low	Middle	High
West Java	−16,936	−54,900	−3,989	−1,597	−11,111	−3,926
Central Java	−22,150	−20,092	−9,288	−1,684	−8,492	−1,647
East Java	−16,657	−13,831	−1,868	−1,420	−9,227	−2,174
South Sulawesi	−5,222	−9,056	−1,061	−413	−2,515	−595
West Sumatra	−948	−8,453	−615	−201	−704	−533
Indonesia	−107,880	−187,334	−15,874	−6,116	−60,190	−26,465

Change in producers' surplus (million rupiahs)

	Rural			Urban		
Region/country	Low	Middle	High	Low	Middle	High
West Java	28,415	37,990	36,548	0	0	0
Central Java	19,033	30,229	30,709	0	0	0
East Java	17,604	32,142	37,834	0	0	0
South Sulawesi	9,142	13,824	8,451	0	0	0
West Sumatra	5,139	6,697	6,088	0	0	0
Indonesia	107,708	182,735	169,387	0	0	0

Net transfer of income (million rupiahs, constant 1988 prices)

	Rural			Urban		
Region/country	Low	Middle	High	Low	Middle	High
West Java	11,479	−16,911	32,559	−1,597	−11,111	−3,926
Central Java	−3,117	10,137	21,421	−1,684	−8,492	−1,647
East Java	946	18,311	35,966	−1,420	−9,227	−2,174
South Sulawesi	3,920	4,767	7,390	−413	−2,515	−595
West Sumatra	4,191	−1,756	5,473	−201	−704	−533
Indonesia	−173	−4,599	153,513	−6,116	−60,190	−26,465

SOURCE: Author's calculations.
NOTE: Assumes 5 percent price change.

Appendix 8.2. The Transmission of Output Price Changes to the Wage Rate

A simple two-sector model demonstrates the importance of the employment share and the elasticity of demand for the transmission of output price changes to the wage rate. The two sectors are denoted with subscripts r (rice) and n (nonrice). When the total supply of labor is fixed, any labor that is attracted to the rice sector as a consequence of an increase in the price of rice must be given up by the nonrice sector:

$$dL_r = - dL_n \qquad (1)$$

where L = total employment in the sector.

If the labor market operates under competitive conditions, the wage rate offered by employers is equal to the value of labor's marginal product. In this case, the increase in the price of rice would create an immediate equiproportional increase in the wage rate. If the labor market is integrated, however, this initial increase cannot be sustained. When wages in the rice sector are higher than wages in the nonrice sector, workers would migrate from the nonrice sector to the rice sector. This movement, which reduces wages in the rice sector and increases them in the nonrice sector, would continue until wages in the two sectors were equal once again.

The change in labor demand in each of the sectors can be measured as the product of three terms—the initial size of the labor force, the elasticity of demand for labor with respect to the wage rate, and the percentage change in the wage rate. Given the assumed price increase for rice, the rice sector would expand and the nonrice sector would contract.

$$dL_r = L_r * dw_r/w_r * E_r \quad (2)$$

$$dL_n = -L_n * dw_n/w_n * E_n \quad (3)$$

where dw/w = the proportional change in the wage rate, and
E = the elasticity of demand for labor $(dL_n/L_r \div dw_r/w_r)$.

Substituting equations (2) and (3) into equation (1) and rearranging terms gives the following expression for the change in the wage rate in the nonrice sector:

$$dw_n/w_n = (L_r/L_n) * (E_r/E_n) * (dw_r/w_r) \quad (4)$$

To focus on the role of the rice price change, an alternative expression can be substituted for the change in the wage rate in the rice sector. The ultimate change in the wage rate in the rice sector can be expressed by manipulation of the marginal value product formula. Total differentiation shows the proportional change in the wage rate, dw/w, as a composite of changes in the price of rice and in the marginal physical product of labor:

$$dw/w = [(dP * MPP) + (P * dMPP) + (dP * dMPP)]/(P * MPP) \quad (5)$$

where MPP = the marginal physical product, and
P = the price of rice.

Substituting equation (5) into equation (4) and ignoring the term dP * dMPP (as second-order small) yields the following expression for the transmission of the rice price to the wage rate:

$$dw_n/w_n = (L_r/L_n) * (E_r/E_n) * [(dP/P) + (dMPP/MPP)] \qquad (6)$$

This expression shows that the larger rice employment is relative to nonrice employment, the larger the transmission of the price change is to the wage rate. The relative magnitudes of the elasticities of demand matter as well, unless they are equal. In the latter case, the elasticity term becomes equal to one, and the transmission of price depends entirely on the relative magnitude of rice employment.

Finally, the relative change in the marginal physical product of labor in rice production is relevant as well. This term reflects the independent influence on the wage rate of the demand elasticity for rice. If the marginal physical product does not decline rapidly in response to increased employment of labor in rice (if the production technology exhibits constant returns to scale, for example), the demand for labor will be relatively elastic. In this event, much labor can be absorbed from the nonrice sector without requiring a very substantial decline in the wage rate. Rapid declines in the marginal physical product, on the other hand, are associated with inelastic demand for labor from the rice sector, and only a small migration of labor from the nonrice sector is necessary to force the wage rate toward the level that prevailed before the rice price increase.

9. Summary and Conclusions

Scott Pearson

This book sets out three alternative targets for Indonesian rice production: high (for output growth of about 4 percent per year), low (about 1 percent), and medium (about 2.5 percent). The study documents what policies would be needed to reach these targets and what the likely consequences of those policies would be for incomes and employment in rural Indonesia. Each target represents a rice strategy involving an interlinked set of policies with broad effects throughout the economy. The likely effects of each strategy are discussed in light of evidence from new surveys of rice production and the rural labor market.

Under the high-growth strategy, Indonesia would produce more rice than it consumes every year. In most years, this would result in a surplus that would have to be exported with subsidies. With the low-growth target, the country would gradually reduce its dependence on rice production, diversifying the rural economy into other crops and nonagricultural activities. Under the medium-growth scenario, Indonesia would maintain long-term self-sufficiency on trend, with production equaling consumption on average, and BULOG would use storage, trade, or both to stabilize rice prices from year to year.

The Past

A price policy environment leading to high profits and stable prices has induced substantial increases in rice production through the application of more variable inputs and on-farm investment in land preparation and other productive assets. In addition, government investment in irrigation and other rural infrastructure has allowed farmers to take full advantage of the latest rice varieties, and research and extension ensured that new techniques continued to become available. The global environment for Indonesian rice policy has been highly unstable so that Indonesian dependence on international rice trade entailed unusually large risks for the government and for rice producers and consumers. Furthermore, Indonesian imports have tended to raise the world price and exports to lower it, especially in the short run. This implies that any policies resulting in dependence on world rice trade—consistently either exporting or importing rice—are risky. To help stabilize domestic prices with lower costs of domestic procurement and storage, however, Indonesia could trade small amounts intermittently.

Detailed new data for current rice production systems in Java, South Sulawesi, and West Sumatra show that rice is overwhelmingly the most profitable crop in these areas. Hence price policy adjustments are unlikely to induce many farmers to switch into or out of rice. A wide range of production techniques now coexist, following the underlying diversity in parcel size, topography, and agronomic conditions. Farmers alter the level and combination of inputs as their conditions change and new techniques become available. Input use, yields, and production levels, therefore, will be affected by the availability of irrigation and other infrastructure, along with appropriate technological packages for each type of farm.

The links among rice production levels, total employment, and wages in rural Indonesia are analyzed with new survey data from three sites on Java and one in South Sulawesi. These data show that changes in the demand for labor directly in rice production are no longer the principal influence on rural wage and employment levels. Instead, the rice sector contributes only a part, albeit an important part, of the total demand for labor. Rural incomes, which determine the demand for transportation, construction, small industry, and other labor-intensive rural activities, combine with urban wages and employment levels to set the demand for and return to unskilled labor in an integrated, competitive labor market. In the 1970s and early 1980s, steady increases in national income raised demand for labor in a wide range of industrial and service activities. These changes helped real rural wage rates to rise in the first half of the 1980s, narrowing the gap between high- and low-wage areas. This increase in labor costs, in turn, has led to the gradual shift away from traditional labor-hiring practices to the increasing use of contract labor and payment of wages in cash.

Rising real wages and other influences induced changes in rice technology, rice production levels, and labor use in rice during the 1970s and 1980s. New technologies have raised rice output and farm incomes without changing the total quantity of labor used. This result provides further evidence that increases in rural and urban income, not increases in the amounts of labor used for rice production, have been the dominant source of higher employment levels in rural areas. Moreover, the spread of synchronized planting and harvesting to meet irrigation schedules or to allow integrated pest management strongly increase the seasonality of labor demand, thus enhancing the attractiveness of mechanization in many areas. This change, which reduces total employment in rice farming, is already occurring in West Java and South Sulawesi.

The Near Future

Forward-looking simulations, using the evidence given above, can indicate what policies would be needed to reach alternative rice production targets, each one defining a separate rice strategy. These simulations quantify the increases in rice area, yields, and total output that can be expected from each policy instrument and show how those instruments fit together to form coherent strategies around each target level of growth.

A substantial burst of rice production, raising annual production by about 1 percent a year for five years, could be achieved through a gradual phaseout of the TRI sugar program, releasing about 30,000 hectares of high-productivity land into rice every year for five years. This deregulation also would cause a direct increase in employment levels because employment in rice production and processing is higher than in sugar.

Continuing high levels of investments in irrigation and other rural infrastructure are needed to complement increases in rice output from the conversion of sugar lands. Although disaggregated data to show precisely which lands could be upgraded most easily to higher-quality irrigation systems are not now available, a number of irrigation investments are under way or planned for REPELITA V. Relatively large increases in rice output can be achieved from either the rehabilitation of existing reservoirs and canals or the construction of entirely new systems. More investment in irrigation, along with continued investment in research and extension services and improved pest management, together can add significantly to growth of rice production. This investment could bring annual rates of output growth from these sources to over 1 percent during REPELITA V (1989–93).

To avoid falling below trend increases in consumption, both changes in regulatory policy (to bring sugar lands into rice cultivation) and increases in public investments (to increase the yields and cropping intensity of poorly- irrigated land) will be needed. If these are not undertaken, Indonesia would likely face either high-cost dependence on imported rice or increases in domestic rice prices. A 5 percent increase in the real price of rice could add perhaps another half percentage point of annual output growth. These enhanced price incentives almost certainly would impose much higher costs on the government and consumers than would changing the regulatory policy for sugar or increasing public investment in irrigation and research. But additional price incentives appear to be necessary to maintain trend self-sufficiency during 1991–93.

If a phaseout of the TRI sugar program on Java, investments in irrigation and other infrastructure, and a 5 percent increase in the rice price in 1991 all occur, it should be possible to maintain trend self-sufficiency during the REPELITA V period. To produce a regular rice surplus, however, the amounts of investment or rice price hikes would need to be much higher and are unlikely to be feasible (given expected levels of government revenues) or desirable (given expected export prices relative to Indonesian policy prices). Consequently, the strategy of trend self-sufficiency, with a target growth rate of 2.5 percent per year, would be the lowest-cost, lowest-risk option.

The equity implications of adopting that strategy are tested by investigating the direct and indirect effects of the major policy instruments on rural incomes, employment, and wages. The new survey data on rural labor markets assist analysis of the effects of price policy, deregulation of sugar land, and investments in irrigation on the key parameters of the labor market in the major rice-producing regions. The results demonstrate that price policy has a significant potential to shift rural incomes because it affects the price of rice for all producers (thus changing their profits) and all consumers (thus influencing their real incomes).

Raising the rice price by 5 percent shifts income from urban centers into rural areas that are net producers of rice. But it also reduces the income of urban or rural residents who are net rice consumers, including most of the very poor. This direct effect would worsen income distribution in all urban centers and in some rural regions. The majority of net consumers of rice would experience about a 1 percent decline in purchasing power if the rice price were increased by 5 percent. Deregulation of sugar lands and investment in irrigation, however, would have a positive effect on rural income and on income distribution. Both of these policies would have a large impact on employment levels because they would increase the intensities of labor and land use.

The estimated net income in rice production generated by the proposed rice strategy during the 1989–93 planning period is Rp 421 billion, about 10 percent of total income earned in rice farming on Java in 1987. Over half of this incremental income derives from irrigation investments, one-third from the phaseout of the TRI sugar policy, and one-eighth from price policy. The estimated net gains in rice employment throughout the five-year projection are 1.1 million labor years—about 15 percent of current rice employment and 3 to 5 percent of total unskilled jobs. About half of these new jobs would be associated with irrigation investment, one-third with price policy, and one-eighth with sugar policy. This estimate is overstated because it does not account for employment losses from reduction of crops replaced by rice or from substitution of purchased inputs for labor in rice farming.

The negative equity impact on rice consumers would be tempered by two offsetting influences, both affecting the wage rate for unskilled labor. The demand by the rice sector for additional unskilled labor would have a direct upward influence on the wage rate; since rice production accounts for 20 to 30 percent of total unskilled employment, this transmission effect could be as much as 1 or 2 percent of the wage rate. In addition, much of the increased returns to labor and land employed in rice production would be expended on consumption or investment, creating more indirect employment of unskilled labor in construction, transport, and other rural services. The likely magnitudes of these offsetting influences are not estimable with existing information. But they might compensate for a significant portion of the negative direct effects of the price change because most poor adult rice consumers are also unskilled laborers.

The Principal Empirical Results

The new data and analysis in this book provide clear evidence that the lowest-cost and lowest-risk rice strategy would be to aim for trend self-sufficiency at about 2.5 percent annual growth in production, achieved through a combination of regulatory reforms producing short-term gains, increases in production from higher public investment in irrigation and other infrastructure, and small increases in the real price of rice. Indonesia has a strong present comparative advantage in producing rice to prevent imports and the likelihood of maintaining a future comparative advantage in rice production.

Planning for a rice production growth rate of less than 2.5 percent annually is unwise because too much reliance would be placed on rice imports in an unstable international rice market. A low-growth strategy of 1 percent annually thus is inconsistent with the objective of national food security (and price stability). At the other extreme, a production growth target of 4 percent annually could be extremely expensive and may not be technically feasible. A high-growth target is inconsistent with the objective of efficient resource allocation because the higher domestic rice prices required for its achievement would likely create the need for export subsidies in most years.

Results from the field surveys show that rice cultivation is highly profitable in both private and social terms. High levels of private profitability mean that policies designed to increase rice production have furthered the objective of rural income growth. Positive levels of social profitability imply that these policies also have been successful in allocating resources efficiently on an economy-wide basis. Relatively high profits in rice indicate that farmers are not likely to move out of rice into other food crops in the near term. In fact, technical constraints prohibit them from doing so in the rainy season. High levels of profitability also indicate that the indirect income and employment effects associated with expenditures and investments of rice profits are important for rural economic growth.

Although rice production has grown rapidly during the past twenty years, aggregate employment in the rice sector has remained stable. Higher levels of employment stemming from rising yields, greater cropping intensities, and expansion of production area have been offset by changes in cultivation techniques (such as the shift from the *ani-ani* to the sickle for harvesting) and the gradual mechanization of certain tasks (such as land preparation and threshing).

Even though demand for labor in rice production in the aggregate has remained stable, real wages for unskilled labor in the rice sector rose rapidly in the first half of the 1980s. Hence policies designed to promote rice production—in the context of rapid growth of the entire Indonesian economy—generally have been consistent with the government's objective of desirable income distribution. Growth in real wages for rural unskilled workers is a principal indicator of successful economic development for Indonesia.

Results from the field surveys indicate that the rural labor market is reasonably well integrated on Java. Unskilled workers face a wide range of job opportunities in both the agricultural and nonagricultural sectors of the economy and allocate their time between activities in response to differences in relative wages. Rural economic growth associated with a dynamic rice economy and higher levels of labor demand in rural and urban areas has improved the opportunities available to unskilled workers in the 1980s. Consequently, national income growth, rather than growth in rice production per se, is likely to be the main source of expanded job opportunities for rural unskilled laborers in the future.

Complementarities and Conflicts among Objectives and Policy Instruments

Three basic indicators of national welfare—aggregate income, income distribution, and food security—have been suggested as fundamental objectives of government. These basic objectives often are in conflict with each other. Some instruments, notably price policy, create trade-offs among these objectives; for example, consumers, who buy more rice than they sell, are harmed by decisions to raise rice prices. But in general, the situation of rice in Indonesia has presented an exceptional, and very fortunate, case in which the three objectives of government policy have been served by a strategy of rice self-sufficiency.

This fortunate situation could change soon. An emerging trade-off between the objectives of food security and efficient income growth is likely to be encountered if Indonesia pursues a strategy of rice self-sufficiency on trend. Increases in the rice price, needed to maintain self-sufficiency, could push domestic prices above the long-term c.i.f. trend and lead to inefficient production. Higher prices also run the risk of causing inequitable income distributional effects for rice consumers. But any future tradeoff between food security and equity would be lessened to the extent that higher rice prices were transmitted into higher unskilled wages and that increased rice incomes were expended on labor-intensive local goods and services.

Along with consumers' resistance to higher prices, the Indonesian government faces other important obstacles in implementing policies to further its objectives. For example, the least expensive source of short-term increases in rice production is the deregulation of land use. This would result in shifts from sugar to rice production, providing higher profits and faster growth of national income, higher employment levels and better income distribution, and more food security. But it also would reduce the use of capacity and thus raise unit costs in Indonesia's large sugar mills, and it would induce greater sugar imports. Some mills probably would have to be shut, which would result in substantial conflict and controversy.

Investments in upgrading the quality of irrigation can have similar effects at the national level—raising aggregate income, creating employment, and improving food security. These investments expanded rapidly with the increasing government income from oil in the 1970s, multiplying the benefits of that income many times over. But the constituency for rural investment is relatively diffuse. Despite its value for the nation as a whole, irrigation has fewer vocal advocates than do many other budgetary priorities. Hence irrigation expenditures fell even faster than other government spending when oil income declined in the 1980s, before recovering near the end of the decade.

A critical policy decision is whether to use only domestic procurement and storage to stabilize prices, tying up capital in large food security stocks, or to supplement domestic stocks with frequent small amounts of imports and exports whenever prices seem favorable given local supply-demand balances. It is almost certain that the stocks-only policy (absolute self-sufficiency) would require higher rice prices, larger government subsidies, and more risk than the stocks-and-trade policy (trend self-sufficiency), but this is an important topic for future research.

The issues for future rice policy are abundantly clear. As this book documents, if Indonesia is to avoid dependence on regular rice imports in the 1990s, it will be necessary to make a concerted effort to maintain prices that will provide positive incentives for farmers (while not placing unreasonable demands on consumers to adjust their food consumption patterns), to deregulate land use as much as possible, and to invest efficiently in irrigation and other rural infrastructure (and its maintenance) and in research for new rice varieties.

How might policymakers best implement a strategy of trend self-sufficiency for rice in Indonesia? Medium-term (five-year) planning calls for rice production gains of about 2.5 percent annually—enough to match the expected growth of rice consumption. During the REPELITA V period (1989–93), two likely sources of annual rice output growth are irrigation investments (about 1 percent) and *gabah* price policy (about 0.5 percent). Because the impact of past public and private investment in irrigation, input distribution facilities, high-yielding seeds, and improved farming practices (the so-called trend effects) is included in the estimates of irrigation and rice price effects, a growth rate of 2.5 percent per year could be achieved only by deregulating TRI sugar land. This desirable policy change could add a full percentage point to growth in annual rice output for five years. Moreover, the deregulation of sugar land would further all three food policy objectives because it would result in an efficient growth of income, improve real incomes and employment, and enhance food security.

Short-term (year-to-year) planning for trend self-sufficiency calls for a flexible use of trade policy to balance unexpected domestic shortfalls or surpluses. Public storage of rice stocks in excess of levels needed for buffer stocking or for maintaining the food security stock (about 1 million metric tons) is generally more expensive than exporting, and public inventories maintained at levels so high as to avoid any imports even in severe drought years are far costlier than rice imports. With this knowledge in hand, Indonesian policymakers have exported surplus stocks at prices far below domestic costs (in 1985–86) and have supplemented domestic supplies with foreign procurement by recalling earlier loans of rice made to the Philippines and Vietnam (in 1988–89). Following an exceedingly good production year in 1989, Indonesia faced the happy prospect of having adequate domestic supplies both to meet all consumption needs and to rebuild food reserve stocks to desired levels. In this circumstance, Indonesia should be able to be self-sufficient in rice on average, while exporting unwanted surpluses or importing to fill temporary deficits.

With this strategy of trend self-sufficiency in place, the Indonesian government could achieve increases in rice output that would improve food security by reducing or eliminating regular imports and generate continued growth in rural incomes with only a minor negative impact on the distribution of that income. But if these priorities are eclipsed by others, future food crises may be needed to bring rice production back into prominence.

References

Adams, F. Gerald., and Jere R. Behrman. 1976. *Econometric Models of World Agricultural Commodity Markets*. Cambridge, Mass.: Ballinger.

Adiningsih, J. Sri, et al. 1989. "The Status of N, P, K, and S of Lowland Rice Soils in Java." Jakarta: Center for Soil Research, Agency for Agricultural Research and Development, Ministry of Agriculture.

Affif, Saleh, Walter P. Falcon, and C. Peter Timmer. 1980. "Elements of Food and Nutrition Policy in Indonesia." In Gustav Papenek, ed., *The Economy of Indonesia*. New York: Praeger.

Affif, Saleh, and C. Peter Timmer. 1971. "Rice Policy in Indonesia." *Food Research Institute Studies in Agricultural Economics, Trade, and Development* 10(26): 121–59.

Altemeier, Klaus, et al. 1988. "Supply Parameters for Indonesian Agricultural Policy Analysis." *Economics and Finance in Indonesia*. 36(1): 47–69.

Bardhan, Pranab K. 1984. "Determinants of Supply and Demand for Labor in a Poor Agrarian Economy: An Analysis of Household Survey Data from Rural West Bengal." In Hans Binswanger and Mark Rosenzweig, eds., *Contractual Arrangements, Employment, and Wages in Rural Labor Markets in Asia*. New Haven: Yale University Press.

Barker, Randolph, Robert W. Herdt, and Beth Rose. 1985. *The Rice Economy of Asia*. Washington, D.C.: Resources for the Future.

Binswanger, Hans. 1978. *The Economics of Tractors in South Asia: An Analytical Review*. New York: Agricultural Development Council.

Biro Pusat Statistik (BPS). 1983. *Agricultural Census*. Jakarta: Biro Pusat Statistik.

———. 1983. [Survey of agricultural marketing]. *Survei Alat-alat Pertanian* Jakarta: Biro Pusat Statistik.

———. 1986. *Struktur Ongkos Dalam Produkis Padi dan Palawija* [Cost of production for paddy and palawija]. Jakarta: Biro Pusat Statistik.

———. 1987 and various years. *Statistical Yearbook*. Jakarta: Biro Pusat Statistik.

Booth, Anne. 1988. *Agricultural Development in Indonesia*. Sydney: Allen and Unwin.

Collier, William 1988. "A Preliminary Study of Employment Trends in Lowland Javanese Villages." Jakarta. Mimeo.

Collier, William, and Achmad Birowo. 1973. "Comparison of Input Use and Yields of Various Rice Varieties by Large Farmers and Representative Farmers." Agro-Economic Survey. Bogor, Indonesia.

Collier, William, et al. 1974. "Agricultural Technology and Institutional Change in Java." *Food Research Institute Studies* 8(2): 169–94.

———. 1981. "Labor Absorption in Javanese Rice Cultivation." Background paper for Technical Meeting on Labor Absorption in Agriculture, June 11–13, Bogor, Indonesia.

———. 1982a. "Acceleration of Rural Development of Java." *Bulletin of Indonesian Economic Studies.* 18(3): 84–101.

———. 1982b. "The Acceleration of Rural Development on Java: From Village Studies to a National Perspective." *Agro Economic Survey,* Occasional Paper No. 6. Bogor, Indonesia.

———. 1988. "A Preliminary Study of Employment Trends in Lowland Javanese Villages" (draft). Jakarta.

Falcon, Walter P., William O. Jones, Scott R. Pearson, et al. 1984. *The Cassava Economy of Java.* Stanford, California.: Stanford University Press.

Falcon, Walter P., and Eric A. Monke. 1979–80. "International Trade in Rice." *Food Research Institute Studies* 17(3): 271–306

Falcon, Walter P., et al. 1985. "Rice Policy in Indonesia, 1985–1990: The Problems of Success." Jakarta: BULOG.

Food and Agriculture Organization of the United Nations, Committee on Commodity Problems, Intergovernmental Group on Rice. 1983. "Tariff and Nontariff Barriers in Rice Trade." CCP:RI 83/5. Rome: FAO.

Forbes, Dean K. 1981. "Mobility and Uneven Development in Indonesia: A Critique of Explanations of Migration and Circular Migration." In Gavin W. Jones and Hazel V. Richter, eds., *Population Mobility and Development: Southeast Asia and the Pacific.* Canberra: Australian National University.

Geertz, Clifford. 1963. *Agricultural Involution: The Process of Ecological Change in Indonesia.* Berkeley and Los Angeles: University of California Press.

Godfrey, Martin. 1987. "The Limits of a Labor Force Approach to Analysis of Indonesia's Employment Problem." Development Studies Project Memo No. 4 Jakarta.

Grant, Warren R., T. Mullins, and W. F. Morrison. 1975. *World Rice Study: Disappearance, Production and Rice Relationships Used to Develop the Model.* USDA Economic Research Service Bulletin 608. Washington, D.C.: USDA.

Hafsah, J., and Richard H. Bernsten. 1983. "Economic, Technical, and Social Aspects of Tractor Operation and Use in South Sulawesi, Indonesia." In *Consequences of Small-Farm Mechanization.* Los Baños, Philippines: International Rice Research Institute and Agricultural Development Council.

Hart, Gillian P. 1986a. "Exclusionary Labor Arrangements: Interpreting Evidence on Employment Trends in Rural Java." *Journal of Development Studies* 22(4): 681–96.

———. 1986b. *Power, Labor and Livelihood: Processes of Change in Rural Java.* Berkeley and Los Angeles: University of California Press.

Hart, Gillian P., and Daniel G. Sisler. 1978. "Aspects of Rural Labor Market Operation: A Javanese Case Study." *American Journal of Agricultural Economics.* 60(5): 821–26.

Hayami, Yujiro, and Anwar Hafid. 1979. "Rice Harvesting and Welfare in Rural Java." *Bulletin of Indonesian Economic Studies,* 15(2): 94–112.

Hayami, Yujiro, and Masao Kikuchi, 1981. *Asian Village Economy at the Crossroads.* Tokyo: University of Tokyo Press.

Hazell, Peter B. R., and Alisa Roell. 1983. "Rural Growth Linkages: Household Expenditure Patterns in Malaysia and Nigeria." Research Report no. 41, International Food Policy Research Institute, Washington, D.C.

Hugo, Graeme J. 1981. "Road Transport, Population Mobility, and Development in Indonesia." In Gavin W. Jones and H. V. Richter, eds., *Population Mobility and Development: Southeast Asia and the Pacific.* Canberra: Australian National University.

Jatileksono, Tumari. 1989. "Impact of Modern Rice Technology on Land Prices across Production Environments in Lampung, Indonesia." Paper presented at the Third Workshop on Differential Impact of Modern Rice Technology on Favorable and Unfavorable Production Environments, RF/IRRI, Dacca, March 29–April 6, 1989.

Kasryno, Faisal, ed. 1984. *Prospek Pembangunan Ekonomi Pedesaan Indonesia* [Prospects for economic development in rural Indonesia]. Jakarta: Yayasan Obor Indonesia.

Korns, Alexander. 1988. "Wage Data at BPS." Development Studies Project II. Memo prepared for BAPPENAS and BPS, February 3.

Lingard, John, and Al Sri Bagyo. 1983. "The Impact of Agricultural Mechanization on Production and Employment in Rice Areas of West Java." *Bulletin of Indonesian Economic Studies* 19(1): 53–67.

Lluch, Constantino, and Dipak Mazumdar. 1985. "Indonesia: Wages and Employment." Washington, D.C.: World Bank.

Maamum, Yusef, et al. 1983. "Consequences of Small Rice Farm Mechanization in South Sulawesi: A Summary of Preliminary Analyses." In *Consequences of Small-Farm Mechanization.* Los Baños, Philippines: International Rice Research Institute and Agricultural Development Council.

Manning, Christopher. 1986a. "Changing Occupational Structure, Urban Work and Landowning Class in Rural West Java." Discussion Paper No. 14., Center for Development Studies. South Bedford Park, Australia: Flinders University.

———. 1986b. "The Green Revolution, Employment and Economic Change in Rural Java: A Reassessment of Trends during the Suharto Era." South Bedford Park, Australia: Flinders University.

———. 1988. "Rural Employment Creation in Java: Lessons from the Green Revolution and Oil Boom." *Population and Development Review* 14(1): 47–80.

Manning, Christopher, and Gunawan Wiradi. 1984. "Land Ownership, Tenancy, and Sources of Household Income." Rural Dynamics Series No. 29. Bogor, Indonesia.

Manwan, Ibrahim, and Achmad Fagi. 1989. "N, P, K, and S Fertilization for Food Crops: Present Status and Future Challenges." Paper presented at the Workshop on Sulfur Fertilizer Policy, Ministry of Agriculture, July 18–20, Jakarta.

Mazumdar, Dapik, and M. Husein Sawit. 1986. "Trends in Rural Wages, West Java, 1977–1983." *Bulletin of Indonesian Economic Studies,* 22(3): 93–105.

Mears, Leon A. 1959. *Rice Marketing in the Republic of Indonesia.* Jakarta: University of Indonesia.

———. 1981. *The New Rice Economy of Indonesia.* Special edition for Badan Urusan Logistik. Jogyakarta, Indonesia: Gajah Mada University Press.

Mellor, John W. 1984. "Food Price Policy and Income Distribution in Low Income Countries." In Carl K. Eicher and John M. Staatz, eds., *Agricultural Development in the Third World.* Baltimore: Johns Hopkins University Press.

Mintoro, Sugianto, and Waluyo. 1984. "Hubungan Kerja Dalam Usahatani Padi" [Work relationships in rice cultivation]. Rural Dynamics Series No. 27. Bogor, Indonesia.

Monke, Eric. 1983. *International Grain Trade, 1950–80.* Technical Bulletin 247. Tucson: University of Arizona Agricultural Experiment Station.

Monke, Eric A., and Scott R. Pearson. 1989. *The Policy Analysis Matrix for Agricultural Development.* Ithaca, N.Y.: Cornell University Press.

Monke, Eric, and Salah A. Salam. 1986. "Trade Policies and Variability in International Grain Markets." *Food Policy* 11(3): 238–52.

Monke, Eric, and Lance Taylor. 1985. "International Trade Constraints and Commodity Market Models: An Application to the Cotton Market." *Review of Economics and Statistics* 67(1): 98–107.

Monteverde, Richard T. 1987. "Food Consumption in Indonesia." Ph.D. dissertation, Harvard University.

Montgomery, Roger D., and Daniel G. Sisler. 1974. *Labour Absorption in Jogjakarta, Indonesia: An Input-Output Study.* Ithaca, N.Y.: Cornell University Press.

Naylor, Rosamund L. 1989. "Rice and the Rural Labor Market in Indonesia." Ph.D. dissertation, Stanford University.

———. 1990. "Wage Trends in Rice Production on Java: 1976–1988." *Bulletin of Indonesian Economic Studies* 26(2): 133–53.

Nelson, Gerald. 1988. "Sugar Policy in Indonesia." Cambridge, Mass.: Harvard Institute for International Development. Mimeo.

Nestel, Barry L. 1987. "Agricultural Research in Indonesia, Its Potential Role in the Development Process with Particular Reference to the Advancement of Biological Technology." Jakarta: Ministry of Agriculture.

Pearson, Scott R. 1990. "Substitutions in End Uses for Food Commodities and Agricultural Trade Policy." *Food Research Institute Studies* 22(1): 109–27.

Petzel, Todd E., and Eric A. Monke. 1979–80. "The Integration of the International Rice Market." *Food Research Institute Studies* 17(3): 307–26.

Rucker, Robert L. 1985. "A Preliminary View of the Employment Problem: Indonesian Option and Realities." Jakarta: U.S. Agency for International Development.

Sawit, Husein, and D. Triono. 1984. "Pola Musiman dan Tingkah Laku Rumah Tangga Buruh Tani Dalam Pasar Tenaga Kerji di Pedesaan Java" [Seasonal patterns and level of household activity of landless laborers in the rural labor market on Java]. Rural Dynamics Series No. 25. Bogor, Indonesia.

Siamwalla, Ammar, and Stephen Haykin. 1983. "The World Rice Market Structure, Conduct and Performance." Research Report 39. Washington, D.C.: International Food Policy Research Institute.

Siamwalla, Ammar, and Alberto Valdes. 1980. "Food Insecurity in Developing Countries." *Food Policy* 5(4): 258–72.

Sicular, Terry, ed. 1989. *Food Price Policy in Asia.* Ithaca, N.Y.: Cornell University Press.

Sinaga, Rudolf. 1978. "Implications of Agricultural Mechanization for Employment and Income Distribution: A Case Study from Indramayu: West Java." *Bulletin of Indonesian Economic Studies* 14(2): 102–21.

Siregar, Masdjidin. 1986. "Factors Underlying Adoption of Rice Power-Threshers in West Sumatra." *Jurnal Agro Ekonomi* 5(2): 1–14.

Sjahrir, Nurmalakartini. 1990. "Workers in the Indonesian Construction Industry: Labor Recruitment and Rural–Urban Migration." Ph.D. dissertation, Boston University.

Soelistyo. 1975. "Creating Employment Opportunities in the Rural Areas of East Java." Yogjakarta, Indonesia. Mimeo.

Squire, Lyn. 1981. *Employment Policy in Developing Countries.* Publication for the World Bank. Oxford: Oxford University Press.

Tabor, Steven, et al. 1988. *Supply and Demand for Foodcrops in Indonesia.* Jakarta: Directorate of Foodcrop Economics and Postharvest Processing, Ministry of Agriculture.

Timmer, C. Peter. 1975. "The Political Economy of Rice in Asia: Indonesia." *Food Research Institute Studies* 14(3): 197–222.

———. 1985. "The Role of Price Policy in Rice Production in Indonesia, 1968–82." Cambridge, Mass.: Harvard Institute for International Development.

———. 1986. "The Role of Price Policy in Rice Production in Indonesia." *Research in Domestic International Agribusiness Management.* 6: 55–106.

———, ed. 1987. *The Corn Economy of Indonesia.* Ithaca, N.Y.: Cornell University Press.

———. 1989. "Indonesia: Transition from Food Importer to Exporter." In Terry Sicular, ed., *Food Price Policy in Asia.* Ithaca, N.Y.: Cornell University Press.

Timmer, C. Peter, Walter P. Falcon, and Scott R. Pearson. 1983. *Food Policy Analysis.* Baltimore: Johns Hopkins University Press.

Varley, Robert C. G. 1989. "Irrigation Issues and Policy in Indonesia, 1986–88." Cambridge, Mass.: Harvard Institute for International Development. Mimeo.

White, Benjamin. 1976. "Population, Involution, and Employment in Rural Java." *Development and Change* 7: 267–90.

———. 1979. "Political Aspects of Poverty, Income Distribution and Their Measurement: Some Examples from Rural Java." *Development and Change* 10(1): 91–114.

———. 1986. "Rural Non-Farm Employment in Java: Recent Developments, Policy Issues and Research Needs" (draft). Jakarta.

World Food Institute. 1986. "World Food Trade and U.S. Agriculture, 1976–1985." Sixth annual edition. Ames: Iowa State University.

Index

Library of Congress Cataloging-in-Publication Data

Rice policy in Indonesia / Scott Pearson . . . [et al.].
p. cm.
Includes bibliographical references and index.
ISBN 0-8014-2524-7
1. Rice trade—Government policy—Indonesia. 2. Food industry and trade—Government policy—Indonesia. 3. Agriculture and state—Indonesia. I. Pearson, Scott R.
HD9066.I62F66 1991
338.1'9598—dc20 90-55751